BUILDING MAINTENANCE MANAGEMENT

THIRD EDITION

Reginald Lee, MPhil, FRICS

BSP PROFESSIONAL BOOKS

OXFORD LONDON EDINBURGH

BOSTON PALO ALTO MELBOURNE

First published in Great Britain by
Crosby Lockwood Staples 1976
Second edition published by
Granada Publishing Ltd 1981
Reprinted with minor updating revisions 1983
Third Edition published by
William Collins Sons & Co. Ltd 1987
Reprinted 1988 by
BSP Professional Books

British Library
Cataloguing in Publication Data
Lee Reginald
    Building maintenance management.—3rd
    ed.
    1. Buildings. Maintenance & repair.
    Management aspects.
    I. Title
    658.2'02

ISBN 0–632–02342–2

BSP Professional Books
A division of Blackwell Scientific
    Publications Ltd
Editorial Offices:
Osney Mead, Oxford OX2 0EL
    (Orders: Tel. 0865 240201)
8 John Street, London WC1N 2ES
23 Ainslie Place, Edinburgh EH3 6AJ
3 Cambridge Center, Suite 208, Cambridge,
    MA 02142, USA
667 Lytton Avenue, Palo Alto, California
    94301, USA
107 Barry Street, Carlton, Victoria 3053,
    Australia

Printed and bound in Great Britain by
Mackays of Chatham PLC, Kent

Contents

Preface

Since the publication of the first edition of this book some ten years ago there has been a growing awareness of the importance of building maintenance in the overall context of property management. Increasingly building owners are beginning to accept that it is not in their best interests to carry out maintenance in a purely reactive manner but that it should be planned and managed as efficiently as any other corporate activity. Inevitably this has placed new demands on maintenance managers requiring them to adopt a more systematic approach to their work and become more involved in the total asset planning process.

For example, their advice might well be sought on such matters as the predicted maintenance and other life-cycle costs for alternative ways of meeting particular building needs so that better informed decisions can be taken on the long term allocation of resources to the related activities of maintenance, refurbishment, new build and replacement. This will lead to a better appreciation of the economic and technical implications of different levels of maintenance expenditure and the true effects of deferring maintenance, the basic objective being to ensure that the right work is carried out, of the right quality, at the right time and at the right price to meet the present and future requirements of the building owner.

The various aspects of maintenance management have been extensively researched, and as was stated in the previous edition, the main problem now lies in the implementation of the findings, Whilst maintenance managers can play an important motivating role, in the last resort the organisational changes needed depend to a large extent on the attitude of building owners to their investment in property and the status accorded to the maintenance function.

This book deals in a systematic and comprehensive way with the approach to the management of building maintenance and gives guidance on the related legislative and other constraints. Generally the text has been updated to reflect new thinking on the subject and sections added to show more clearly the relationship between maintenance and the broader aspects of property management. Changes in legislation have necessitated a number of amendments to the chapter on statutory requirements and, in particular, the main requirements of the Building Act 1984 are described. The section on computers has been extended to cover recent developments and possible applications to maintenance decision making and information feedback. Although a computer by itself cannot turn a bad management system into a good one, it can stimulate change by requiring people to look more critically at what they are doing and to reconsider roles and responsibilities and the basic structure of the maintenance organisation. Clearly it is not possible in a book devoted primarily to the management aspects of maintenance to delve too deeply into technology but it was thought that it would be helpful to readers to include a new chapter on the causes of the more common building defects.

It is hoped that the concepts contained in this book will prove equally useful to practitioners and students alike and will assist in developing a more professional and systematic approach to maintenance management.

The author is indebted to the Controller of Her Majesty's Stationery Office for permission to reproduce from Practice in Property Management, Appendices A and B; to RIBA Publications Ltd for permission to reproduce the Minor Works Agreement (Appendix C); and to the Illuminating Engineering Society to reproduce from their Technical Report No 9 Figure 2.4 and Table 2.1.

R. Lee
1986

CHAPTER ONE

Maintenance system

Economic and social significance

The built environment expresses in physical form the complex social and economic factors which give structure and life to a community. The condition and quality of buildings reflect public pride or indifference, the level of prosperity in the area, social values and behaviour and all the many influences both past and present which combine to give a community its unique character. There can be little doubt that dilapidated and unhealthy buildings in a decaying environment depress the quality of life and contribute in some measure to anti-social behaviour. Unfortunately these social consequences are difficult to quantify and as a result are rarely given proper consideration.

In most cases maintenance decisions are based on expediency and over a period of time represent a series of *ad hoc* and unrelated compromises between the immediate physical needs of the building and the availability of finance. There is a lack of precise knowledge of the benefits which accrue from different levels of maintenance expenditure and little attempt is made to forecast the overall long-term effects of doing or forbearing to do work in this field. The reason may be that from the standpoint of the individual firm the amount spent on maintenance appears small in comparison with other operating costs. But when viewed on a national scale it is quite clear that maintenance is an activity of primary importance. The total value of buildings and works accounts for two-thirds of the nation's capital stock. Not only does this represent wealth accumulated over many years but it is also a vital factor in the production of new wealth. The preservation of the value and utility of the stock of buildings is therefore essential to the economic wellbeing of the country.

However, available evidence would suggest that in general buildings are under-maintained and that a substantial part of the building stock is in danger of deteriorating below the point of economic repair. The English House Condition Survey 1981[1] revealed that of the 18.1m dwellings in England around 2m were in poor condition and of these over 1m required repairs costing in excess of £7,000 (at 1981 prices). Further evidence of the dilapidated state of the housing stock was contained in a report published in 1986 by the Audit Commission on Managing the Crisis in Council Housing which estimated that 85 per cent of the 5m council houses were in need of repair and improvement. This shows little change from the situation reported in 1970 by the National Institute of Economic and Social Research[2] which concluded that the arrears of housing maintenance probably amounted to eight or nine times the volume of work actually carried out. Although there are no comparable figures for non-residential buildings it is probable that they suffer from a similar degree of neglect.

Satisfaction of building needs

In order to put the problem into perspective it is necessary to view maintenance in the context of the overall building process. The building needs of the community are met by the inter-related construction activities of maintaining, modernising and replacing the existing stock of buildings and by the erection of additional new buildings (Fig. 1.1).

A frequently expressed view is that the level of expenditure on maintenance is too high, and no doubt it was concern over the magnitude of the total national expenditure on this activity which prompted the Government to set up the Committee on Building Maintenance in 1965. However, in order to judge the reasonableness of the expenditure it is necessary to consider three factors – firstly, whether the amount spent is excessive in

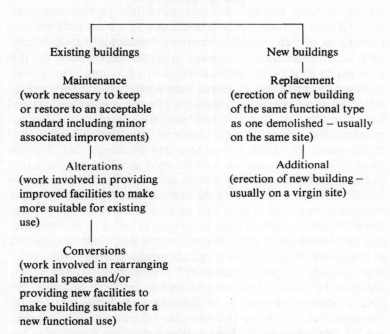

Fig. 1.1. Construction activities within the overall building process

relation to the work done, secondly, whether the work which is done is both necessary and unavoidable, and thirdly, whether it would be advantageous to carry out more work. It would seem that the views expressed are well founded in relation to the first point and that better management and work planning would result in a more economic use of resources and a corresponding reduction in total costs. In regard to the second point there are many reported cases of early maintenance which could have been avoided by better design and the choice of more suitable materials.

The third point concerning the desirable level of maintenance cannot be considered in isolation. Clearly construction resources are limited and the object should be to achieve the optimum allocation of manpower, materials and capital between the maintenance and improvement of existing buildings and the construction of new buildings. Maintenance by arresting decay extends the physical life of a building and thereby delays replacement and defers expenditure on new construction. Of course, it can only be regarded as a substitute for new construction if the building remains functionally satisfactory. To a large extent this will depend upon the degree to which it is possible to modify the internal layout to accommodate changing user requirements. There is thus a relationship between the adaptability of the design, building life, maintenance costs and new construction costs. Generally the more adaptable the building, the longer the period of use, resulting in higher total maintenance costs but lower new construction costs. For the most part, building labour can be used indifferently for either purpose and thus policies which influence the volume of new construction will also affect the availability of labour for maintenance. Stone[3] has argued that more output would be obtained from the same labour force by giving priority to maintenance and improvement, in that this would not only increase the rate at which satisfactory dwellings are made available but would also reduce the cost. Needleman[4] also emphasises the advantages to be gained from rehabilitation rather than wholesale redevelopment; a policy which was implemented by the Housing Act, 1969, by providing much more generous grants for the improvement of houses which lack the basic amenities or which fall below an acceptable standard. However, if policies are adopted which by persuasion, financial inducement or otherwise result in an increase in the demand for maintenance it becomes even more important that this work should be properly planned.

The building process

The role played by maintenance in the overall building process is shown in Fig. 1.2. The extent to which maintenance is considered at the design stage is likely to depend upon whether or not the owner, or the person commissioning the building, will be the subsequent user. Where the owner is a developer who intends selling or leasing the building on completion it is probable that maintenance will be considered only in so far as it is likely to affect the sale price or rent. Even where the owner and user are one and the same, it is probable that the initial

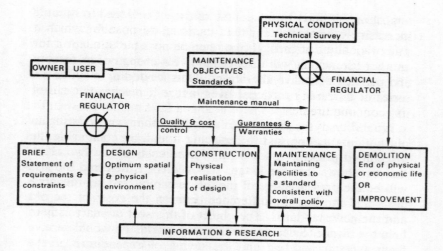

Fig. 1.2. The total building process, showing the role played by maintenance

capital costs of the building and the subsequent maintenance costs will be drawn from different sources and not related. Also it is misleading to think of the owner and user as being individuals, in that for a building of any size they would constitute a complex system of differing interests; often a good deal of abortive work results from failure to recognise and resolve the conflict of interests and objectives. However, both new construction and maintenance are usually subject to financial constraints and, notwithstanding the difficulties, an attempt should be made to achieve a proper balance between the costs of these activities by analysing the costs-in-use of alternative design solutions.[5] Where the design requirements are stated in the form of performance specifications these could be used for establishing maintenance standards and incorporated in a maintenance manual together with other relevant detail concerning the construction of the building and its services.

Eventually a decision will have to be made as to whether to demolish the building and replace it with a new one or to improve or adapt it to make it more suitable for either the present use or a new use. In such cases it is helpful to distinguish between physical life, functional life and economic life. Provided the structure is basically sound it is possible with proper maintenance to extend the physical life of a building almost indefinitely. In the course of time it will tend to become increasingly unsuitable for the functional purpose for which it was originally designed. It may then be possible to adapt it for another use, which will usually involve a change of ownership and perhaps extensive alterations. Thus a building may have a series of different functional lives before it reaches the end of its economic life.

Noble[6] analyses the case for spending money on updating an old building as opposed to replacing it and, by comparing the discounted present values of the future costs of the two alternatives, shows that the longer replacement is deferred the less will be the present value of the future capital and maintenance costs and the less the difference between the costs of the old and the new standards. The object of this sort of analysis is to help the owner to decide whether the benefits he would receive from a better building and the saving of expenditure on the existing building are together worth the cost of replacement. Although it may be difficult to quantify precisely some of the benefits, the analysis nevertheless provides a more rational basis for the replacement decision.

In some cases the economic life may be limited by a substantial rise in the value of the site on which the building stands. This may dictate the replacement of the building with one which will exploit the potential of the site to greater advantage. The determining factor will be the extent to which the value of the replacement building exceeds the cost of building it and the value of the present building. Switzer[7] describes economic life as 'that period of effective life before replacement; replacement will take place when it will increase income or reward absolutely, but will not reduce the rate of interest at present enjoyed from the landlord's total investments'. He advocates the control of the economic life of buildings by means of adaptability in the structure and by the judicious use of planning controls to regulate site values and thereby create the correct conditions for preservation or redevelopment. This would permit a closer relationship to be achieved between design durability and effective life.

Other factors that would have to be considered include social acceptance and the effects of new technology and legislation.

Life cycle costs

Life cycle costs (LCC) are the total costs of owning and using an asset over its predicted life span.

$$LCC = Ic + (Mc + Ec + Cc + Oc) + (VC) - Rv$$

where

Ic	=	initial cost
Mc	=	maintenance costs
Ec	=	energy costs
Cc	=	cleaning costs
Oc	=	overhead and management costs
Vc	=	utilisation costs
Rv	=	resale value

The basic concept is that decisions on the design and acquisition of durable assets should take into account the long-term financial consequences and should not be based solely on initial costs. The LCC method can be used at all stages in the life cycle of a building from inception to eventual sale or demolition:

1. At the inception stage it may be used to determine the most economic way of meeting a need for additional building space. The options to be compared may include:

   (a) Rearrangement of the internal spaces within an existing building.
   (b) Building an extension.
   (c) Gradual redevelopment of the existing site.
   (d) Development of a new site.
   (e) Purchase or lease of another building.

2. During the early design stage of a new building to assist in developing the most economic plan shape, structural form and internal layout. It should be borne in mind that the earlier the LCC technique is applied, the greater the possible savings and the lower the committed costs. Thus the client's requirements should be stated initially in very broad terms so as not to limit the choice of options.

3. During the detail design stage to identify the design features, components and finishings that have the lowest total costs. This stage would culminate in the preparation of

a Life Cycle Cost Plan which should be incorporated in the maintenance manual for the guidance of the property manager.

4. During the occupation of the building to assist in formulating planned maintenance and renewal policies. It also provides a means of identifying high cost areas and evaluating changes that will reduce these costs.

The costs may be broadly divided into Capital and Revenue – a distinction which is usually necessary for accounting and taxation purposes. Unfortunately the levels of these two types of expenditure are generally determined by different decision processes and are subject to different control systems and separately funded. Because of this the significance of total costs is obscured. For LCC to be effective it is essential that decisions on initial capital costs should be tempered by a knowledge of the consequential running costs. One way to ensure that this is done is to require design proposals to be accompanied by an LCC Plan.

Figure 1.3 shows the main life cycle costs and the approximate percentages of initial cost for the various items of revenue expenditure. The percentages are based on an analysis of the life cycle costs of an office block and as will be seen the total revenue costs amount to about 10 per cent of the initial costs.

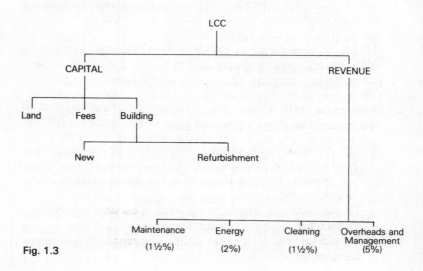

Fig. 1.3

The approach to LCC should follow the pattern illustrated in Fig. 1.4 and itemised below:

1. Establish basic objectives. This should be a clear statement of what the proposals are intended to achieve and may range from the life cycle costs of a new town to the choice of roof coverings for an individual building. For the property manager it could include an evaluation of alterations to an existing building or the financial consequences of renewing rather than repairing a major piece of equipment.
2. Formulate alternative means for achieving the stated objectives.
3. Decide upon a finite planning horizon which is applicable to all the alternatives.
4. Identify all the costs and revenues which are directly relevant to the comparison of the alternatives.
5. Adjust the costs to a common time period by converting to Present Values or Annual Equivalents.
6. Carry out a Sensitivity Analysis to assess the effect of errors in predicting building and component life and different rates of interest.

Flanagan and Norman[12] identify three distinct but closely related aspects of LCC, which are:

1. Life Cycle Cost Planning (LCCP), which focuses on the planning of future costs as well as initial costs.
2. Life Cycle Cost Analysis (LCCA), which involves the systematic collection of running cost data on completed buildings and linking this with the physical, qualitative and performance characteristics of those buildings.
3. Life Cycle Cost Management (LCCM), which covers the application of the techniques to existing buildings and systems and has the following aims:

   (a) To monitor and explain differences between LCCP projections and actual performance.
   (b) To improve the efficiency of the building through more effective utilisation.
   (c) To provide information on asset lives and reliability factors for accounting purposes.
   (d) To aid in the development of an appropriate maintenance policy for the building.

(e) To provide taxation advice on building related items.

Thus LCCM forms an important part of property or asset management.

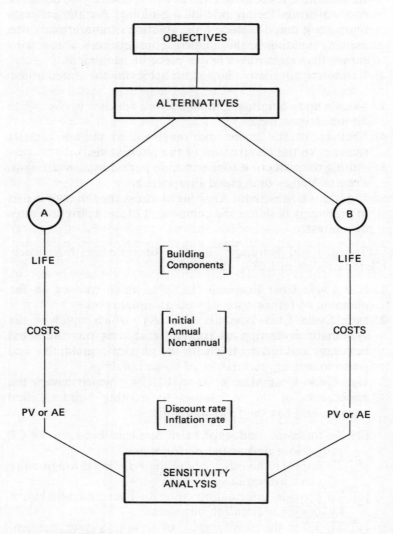

Fig. 1.4. LCC Method

Asset planning/management

In the case of large property owning organisations it is essential to adopt a systematic approach to the overall management of building assets. This will enable a better assessment to be made of the desirable level of maintenance expenditure. Too often funds are allocated purely on the basis of short-term expediency and the long-term penalties of deferring maintenance are ignored. The result is that building deterioration is unchecked and a point is soon reached when disproportionately large sums have to be spent on expensive remedial work. In the meantime the buildings become progressively less efficient and the activities which they house suffer.

The whole process should start at the design stage of each new building with a detailed analysis of the life cycle costs so as to achieve the right balance between initial construction costs and the subsequent maintenance and running costs. The building owner is thus made aware at an early stage of the amount that should be set aside for maintaining the building to an acceptable standard. It should be part of the designer's brief to provide a maintenance manual outlining amongst other things the main elemental life cycles and the future maintenance and energy requirements. Similar life cycle cost analyses should be carried out at regular intervals during the occupation period to take into account changes in the condition and use of the building(s). By this means a strategic asset plan can be drawn up providing for the long-term allocation of resources to the related activities of maintenance, refurbishment, new build and replacement, to reflect the changing building requirements of the organisation.

The following represents a systematic approach to asset planning:

1. Compile a detailed property database.
   The following is the sort of basic data needed:

Location	Space usage
Age	Remaining life
Function	Occupation costs
No. of storeys	Replacement value
Construction	Site value
Floor area	Constraints, e.g. listed building
Element areas	External works
Services	

In the case of new buildings the information may be obtained from 'as built' drawings, bills of quantities suitably amended to allow for variations, or a maintenance manual. For older buildings in the absence of suitable records the necessary data would have to be compiled from inspection reports or special surveys.

2. Determine the condition of the building(s).

For this purpose a condition survey would have to be carried out. This comprises a limited inspection concentrating on those elements at greatest risk and most likely to fail. In order to standardise the approach a check-list of the items to be inspected should be prepared with a simple grading system to record the severity of defects. For large housing estates of a similar type of construction and age it would be sufficient to take a random sample. The object is to give a general picture of the state of the buildings, from which the approximate cost of remedial work can be estimated.

3. Analyse the usage and performance of building spaces.

There is little point in carrying out full-scale maintenance on building spaces which are under-utilised or not utilised at all. If maintenance expenditure is to be justified it must be shown that there is a return of some sort, whether it be financial or merely in terms of comfort and convenience.

The current and possible alternative uses of the building spaces should be considered, taking into account their functional efficiency and the effects of new legislation, new technology and changes in user requirements. In commercial and industrial buildings where requirements are subject to fairly rapid change the regular updating of the premises may be seen as an alternative to maintaining the original facilities. This approach has been described as 'planned obsolescence' and consists of a continuing programme of refurbishment to keep abreast of market trends. It is particularly appropriate to shopping areas where it is necessary to keep ahead of competitors in meeting changes in the shopping pattern of consumers and in retailing methods. Similar periodic assessments of standards are necessary for housing estates to provide for changes in family structure, social and work patterns and tenant preferences. The aim is to consider not only efficiency in satisfying present demands but also changes that will be needed to satisfy future demands.

4. Apply life cycle cost techniques to optimise all resource implications.

This involves quantifying the costs and benefits of alternative strategies to meet the long-term needs of the organisation. For example, in the case of alterations to the internal layout of a commercial or industrial building the benefits to include in the equation may cover such matters as reductions in staff costs, lower running costs, more economic use of production machinery, improved public image, etc. Similar considerations would apply to repair/renew decisions in relation to the expected remaining life cycle of the building.

5. Formulate an investment programme itemising the expenditure requirements of the various activities and stating how the money is to be raised.

The financial plan should itemise all the short- and long-term expenditures required for the proper upkeep and improvement of the buildings so that adequate provision can be made for the necessary funds. This would remove one of the objections that is sometimes made about Repairs Funds, that the expenditure is too uncertain. It would also be of great assistance in the case of buildings in multiple occupation where a service charge is levied on tenants and owners for the maintenance of the common parts and facilities. It would enable tenants to know well in advance what commitments they have for future maintenance costs and allow expenditure on large-scale repairs to be spread over a number of years.

6. Prepare an integrated action plan.

This should set out the timing and resource needs of the operations necessary to ensure that the maintenance policy is part of and complementary to the total programme of maintenance, refurbishment, new build and replacement. It should include such matters as standards of upkeep, priorities, criteria for judging levels of maintenance expenditure, the organisation and decision-taking bases for the maintenance function and the degree of control required through the Management Information System.

Maintenance defined

BS 3811:1964 defines maintenance as 'a combination of any actions carried out to retain an item in, or restore it to, an acceptable condition'. The actions referred to are those associated with initiation, organisation and implementation. There are two processes envisaged: 'retaining', i.e. work carried out in anticipation of failure, and 'restoring', i.e. work carried out after failure. The former is usually referred to as 'preventive maintenance' and the latter as 'corrective maintenance'. There is also the concept of an 'acceptable standard'. This may be construed as acceptability to the person paying for the work, to the person receiving the benefit or to some outside body with responsibility for enforcing minimum standards. It can also be construed more widely as acceptability to the public at large or to specific sections of the public. Clearly there are no absolute standards which would be equally acceptable to everybody or which would remain acceptable to the same group of people over a period of time.

The standards acceptable at the time of undertaking the work may be higher or lower than the initial design standards. In many cases the standard deemed acceptable would be higher than that originally provided and the work would include an element of improvement. It could be argued that this interpretation would be inconsistent with the phrase 'to retain or restore' in that this would suggest the initial standard as the proper basis. However, with the passing of time buildings are modified to accommodate new uses and it becomes increasingly unrealistic to think in terms of keeping or restoring the initial standards.

In recognising the desirability of including a reasonable element of improvement, the Building Maintenance Committee [8] recommended the following definition: 'Work undertaken in order to keep, restore or improve every facility, i.e. every part of a building, its services and surrounds, to a currently accepted standard and to sustain the utility and value of the facility'. This introduces the notion of a 'currently accepted standard' which, from the general tenor of the definition, is assumed to be higher than the initial standard. There may, of course, be cases where buildings are put to a less demanding use for which lower standards would be acceptable. There is also reference to 'utility and value', which would be important factors to take into account when fixing an acceptable standard. The effects on both the value in use and the value on sale or letting would have to be considered. There are difficulties in relating the value on sale to the standard of maintenance, in that the market value of property is determined by many factors other than physical condition. Commercial properties in particular show a tendency to appreciate in value as a result of surrounding development rather than the attributes of the building.

In relation to hospital buildings the Woodbine Parish Report[9] gives the following amplified version of the British Standard definition: 'Work undertaken to keep or restore hospital premises to acceptable standards of safety and efficiency having due regard to the needs of patients and staff within their immediate environment, the requirements of the NHS and the resources available'. The additional words used serve merely to emphasise the constraints on the carrying out of the work. Clearly the standards would be related to safety and efficiency and the work done in such a way as not unnecessarily to disrupt services or inconvenience staff and patients. The reference to available resources is interesting in that it suggests that some arbitrary sum of money is set aside for maintenance and that this cannot be exceeded even though to achieve an acceptable standard would involve a greater expenditure. Thus the standard is really determined by the amount of money allocated rather than as a result of assessing the benefits obtained from maintaining the building to a particular state. The Report goes on to say that 'most items of maintenance should have an element of improvement so as to secure the benefits of new knowledge, products and materials'.

A more functional definition proposed by White[10] is that 'maintenance is synonymous with controlling the condition of a building so that its pattern lies within specified regions'. The word 'control' suggests a positive activity which is planned so as to achieve a defined end result. The term 'specified regions' presumably has a meaning similar to 'acceptable standards' and would be determined in a similar way. An interesting aspect of the definition is that it envisages a range of acceptability with upper and lower limits between which the condition of the building must be maintained.

Maintenance or improvement

The problem of distinguishing between maintenance and improvement is not solved satisfactorily by any of the aforementioned definitions. It is generally conceded that maintenance should include a reasonable element of improvement, e.g. the replacement of worn-out components with up-to-date versions. As a result of technological developments replacements are rarely exactly the same as the original, being superior in some respects but perhaps inferior in others. For instance, a replacement boiler may have a better technical performance but a shorter life than the original. In such cases it is difficult to assess the notional element of improvement, especially where the modern component, as a result of improved manufacturing processes, is cheaper than the original. It is logical therefore to extend the meaning of maintenance to cover localised improvements of this sort. But where the intention is to increase the efficiency in the use of the building by adding facilities which were not previously present, the work should be classed as improvement.

The distinction between maintenance and improvement assumes greater importance when viewed in the context of feedback to the designer. There is a tendency to believe that the level of expenditure on maintenance would be significantly reduced by better-informed initial design. In order to test the validity of this belief, it is necessary to distinguish between work rendered necessary by technical design faults and that directed to improvements for purposes which could not have been foreseen by the designer and incorporated in the original scheme. The usefulness of the information fed back to the designer will therefore depend upon the extent to which it distinguishes:

1. Work which may be regarded as normal in relation to the constructional materials and conditions of use. For instance, where the designer has made a conscious decision to use relatively short-lived materials for a particular purpose, the resulting high maintenance costs would be a normal consequence of that decision.
2. Work resulting from design faults in relation to either
   (a) technical errors concerning inappropriate constructional details or materials;

(b) errors of layout in relation to size, arrangement and juxtaposition of working spaces.

3. Work made necessary by a demand for higher standards or a change in the pattern of use which could not have been foreseen at the time of the initial design.

Although technical design faults attract most attention, it is probable that layout faults prove very much more costly in the long run.

Planned preventive maintenance

BS 3811 defines the different types of maintenance in the following ways:

1. Planned maintenance. Maintenance organised and carried out with forethought, control and the use of records to a predetermined plan.
2. Preventive maintenance. Maintenance carried out at pre-determined intervals or to other prescribed criteria and intended to reduce the likelihood of an item not meeting an acceptable condition.
3. Running maintenance. Maintenance which can be carried out whilst an item is in service.

Preventive maintenance is normally planned and hence the term planned preventive maintenance. It is a concept which is probably more applicable to plant and equipment which is subject to mechanical wear but there are certain building elements which justify this treatment. In order to introduce such a system it is first necessary to produce an inventory of every building, area, service, etc. which has to be maintained. Then it is necessary to determine which items should be included in the planned preventive programme and the frequencies at which they will require attention, e.g. weekly, monthly, quarterly or annually. The selection would be based on the consequences of failure in regard to such factors as safety and productivity, e.g. fire doors would obviously be included, and the frequency on an analysis of past records. Finally job cards are prepared for the various tasks and an appropriate manual or computer bring-forward system devised. Performance should be monitored continuously to check that the work is being carried out in accordance with the programme and that the costs are commensurate with the benefits, e.g., reduction in emergency work and user requests.

Bushell[11] suggests that planned preventive maintenance is worthwhile if:

1. it is cost effective,
2. it is wanted to meet statutory or other legal requirements,
3. it meets a client need from an operating point of view,
4. it will reduce the incidence of running maintenance necessitating requisitions for work from the user,
5. there is a predominant incidence of work for the craftsman rather than pure inspection.

Additionally one could add that the very act of planning, involving as it does an analysis of past performance in order to predict the future, leads to a more enlightened approach to the management of maintenance operations.

Maintenance policies

BS 3811 defines maintenance policy as a strategy within which decisions on maintenance are taken. Alternatively it may be defined as the ground rules for the allocation of resources (men, materials and money) between the alternative types of maintenance action that are available to management. In order to make a rational allocation of resources the benefits of those actions to the organisation as a whole must be identified and related to the costs involved. It is necessary, therefore, to consider the question of policy under the following heads.

1. Objectives. What does maintenance have to achieve? This should be viewed in the context of the organisation's overall building needs. Maintenance is an important part of the terotechnology approach which has been defined as a combination of management, financial, engineering and other practices applied to physical assets in pursuit of economic life-cycle costs. It requires all departments in an organisation to co-operate in ensuring that the assets of the organisation are planned, provided, maintained, operated

and disposed of at the lowest total cost to the organisation. Life-cycle costing, or costs-in-use, is a technique which is usually thought of in terms of initial design decisions but it is equally useful for appraising expenditure on maintenance, alterations and improvements during the life of a building.

2. Benefits. What is to be gained? The benefits may be either short-term or long-term and may be classified as financial, technical or human. The financial benefits spring from a more effective use of the building and are reflected in higher productivity, less wastage of materials, improved sales figures, etc. The technical factors are related to the preservation of the physical characteristics of the building and its services and are reflected in fewer breakdowns with a reduction in downtime and fewer calls for emergency repairs, less accidents, lower future maintenance costs, etc. The human factors are related to the psychological effect of the condition of the building on the user and are reflected in such things as a lower rate of staff turnover with reduced recruiting and training costs, better customer relations and an improved public image. Clearly some of these benefits are difficult to quantify but some attempt should be made to express them in money terms so that the analysis can be as complete as possible.

3. Policies. How shall we proceed? This involves laying down operational and cost objectives for the maintenance department starting with the identification of maintenance tasks, the standards to be achieved and the limits of cost. This will lead to policies concerning the proper balance between preventive and corrective maintenance, how far work should be programmed rather than relying on user requests, the priority to be accorded to different types of work, whether the work is better carried by direct labour or contract and, where the properties are dispersed over a wide area, the extent to which decision taking should be decentralised. These policies will determine the structure of the maintenance organisation and the roles and duties of the supervisory staff.

Maintenance generators

The agencies which act upon the building and erode the initial standards include:

1. Climatic conditions which vary in severity according to the location and orientation of the building and which have the greatest effect on the external elements.
2. User activities including both human and mechanical agencies and authorised and unauthorised usage. For this purpose a burglar may be considered to be a user of the building, albeit an illegal and unwelcome one.
3. Changing standards and tastes which, while not worsening the existing condition, may create a demand for work to be carried out more frequently than functionally necessary; for example, repainting for the sole purpose of changing the colour scheme.

In all these examples, the primary cause may be regarded as either 'normal' or 'abnormal' according to expectations based on past experience. Also, the work required may relate to preserving either the 'functional' or the 'aesthetic' properties of the facility.

The extent to which these agencies cause deterioration and thus create a need for remedial treatment will depend upon:

1. The adequacy of the design and the suitability of the materials specified.
2. The standard of workmanship in the initial construction and subsequent maintenance.
3. The extent to which the designer has anticipated future needs.

Over a period of time the gap between the standards demanded by the user and those provided by the building is likely to widen as shown in Fig 1.5. The diagram is an abstract representation of the building as a whole and quite clearly is not applicable to each and every element, e.g. the requirements for the foundations are unlikely to change and, in the absence of a design error, should last the life of the building without attention. Also, individual elements will vary considerably with regard to the rate at which changes take place and the magnitude of the changes.

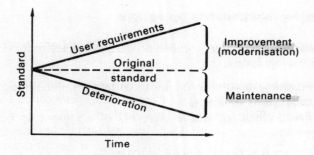

Fig. 1.5. The passage of time related to maintenance requirements

Another factor which is difficult to incorporate in a model of this type is the interdependence of elements and the change in the pattern of deterioration caused by the failure of other elements. Thus an apparently minor defect may have the following consequences:

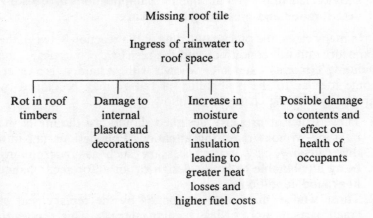

Admittedly this particular chain of events could have been anticipated by the designer, and a layer of felt provided underneath the tiles to convey safely to the gutter any rainwater that might penetrate the covering. However, while this second line of defence would operate satisfactorily for a time, excessive delay in replacing the tile could lead to consequential damage which, although not capable of exact prior evaluation, would clearly be more costly to remedy than the timely replacement of the tile. And, of course, a similar chain of events could be triggered off by other quite different causes, e.g. excessive condensation.

Timing of maintenance operations

The work necessary to combat progressive deterioration takes the following forms:

Patching, involving the more or less regular replacement of small parts or areas.

Replacement of whole elements of components because they:

(*a*) are functionally unsatisfactory;
(*b*) incur high maintenance or running costs;
(*c*) are aesthetically unacceptable.

Preservation of protective coatings either for the purpose of extending the life of the protected material or to maintain appearance.

Cleaning which, although often regarded as a separate activity, has important maintenance implications in arresting deterioration and preserving appearance.

In many cases, the precise outcome of the reaction between the exposure and use conditions and the resisting properties of the building elements cannot be known with certainty. However, some idea as to the probability of failure may be gained by grouping building elements as follows:

1. Those which if properly designed should last the life of the building without requiring attention. Foundations fall into this category, any work necessary such as underpinning being attributable to faulty design or an unforeseen change in ground conditions.
2. Those whose life can be prolonged by the replacement of small parts at more or less regular intervals. This applies to such elements as roof tiling, although a time would come when it would be more economic and functionally more satisfactory to renew the whole of the coverings.
3. Those components which are subject to wear through either human or mechanical agencies. Thus the life of floor finishes is related to the density and type of traffic.
4. Those which are prone to obsolescence as a result of technological advances or merely changing fashion. This applies to all types of surface finishes and fittings such as sanitary appliances. One finds, for example, that the appearance of

glazed brick dadoes in older schools is now considered too 'institutional'; in spite of their excellent maintenance qualities, they are often covered with other finishes which have inferior maintenance characteristics.

It is difficult to apply a time scale to changes in taste but it is reasonable to suppose that the rate of change will increase and there will be an increasing tendency to reject components which, while still functionally adequate, do not measure up to the accepted 'norms' of the period.

5. Those which are exposed to the weather and will in the course of time fail or become disfigured to such an extent that repairs or extensive cleaning will be necessary. This category includes roof coverings and external wall claddings and facings of all types.

6. Protective coatings: these require special consideration in that, although the coating may have some aesthetic value, the primary reason for renewal is to extend the life of the protected element. This type of work forms a high proportion of the total maintenance expenditure and methods of optimising renewal cycles will be considered later.

Maintenance cycles

Assuming normal conditions of exposure and use the periodicity of major renewals might follow the pattern indicated in Fig. 1.6. For this purpose a building life of 60 years has been taken and the renewal periods estimated in multiples of five years. In addition to the major renewals and replacements there will be a large number of smaller items which are of uncertain timing but which in total tend to remain fairly constant from year to year.

If it is assumed that these routine maintenance costs amount to 0·5 per cent of the initial cost each year, i.e. 2·5 per cent of initial cost per quinquennium, the overall cash flow over the life of the building will be as shown in Table 1.1. This has been derived from the percentages and renewal frequencies shown in Fig. 1.6. The percentages shown in the second column of Fig. 1.6 are the percentages of the total initial cost attributable to the individual elements for the building on which the analysis is based. Thus the maintenance cost for the first quinquennium is

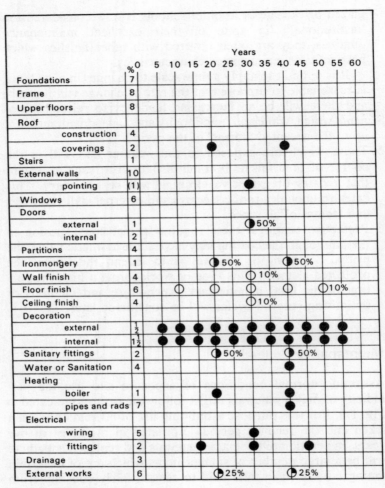

Fig. 1.6. Renewal cycles

made up of 0·5 (external decoration) + 1·5 (internal decoration) + 2·5 (routine) = 4·5 per cent. This is given purely as an example of a method which can be used to assess maintenance costs over the life of the building. Clearly different buildings will have different elemental cost patterns and renewal periods will vary according to design, location and use. In this example the maintenance costs have been expressed as a percentage of

Table 1.1. Cash Flow at Current Prices as percentage of Initial Cost

Year	% initial cost per quinquennium	% initial cost cumulated
5	4·5	4·5
10	5·1	9·6
15	6·5	16·1
20	11·1	27·2
25	4·5	31·7
30	14·4	46·1
35	4·5	50·6
40	20·6	71·2
45	6·5	77·7
50	5·1	82·8
55	4·5	87·3
60	2·5	89·8

the initial cost of the building but, given the initial cost, they could be expressed in monetary terms either for the building as a whole or per 100 m² of floor area.

No attempt has been made to allow for the effects of inflation on future maintenance costs, on the assumption that the funds which bear the expenditure will inflate at the same rate. However, it is unlikely that maintenance costs and new construction costs will continue to bear a constant relationship over a prolonged period of time. Labour forms a higher proportion of the cost of maintenance than of new construction and as wage rates are currently increasing at a faster rate than the prices of most materials it is to be expected that maintenance will become more expensive in relation to new construction.

Also, no account has been taken of the time scale of the payments, i.e. that future disbursements can be met by investing smaller sums now at interest. For instance, £100 invested now at 10 per cent will be worth £110 after one year or, conversely, £110 at the end of the year is worth only £100 now. The present value (PV) of future sums can be calculated from the following formula:

$$\text{PV of } £1 = \frac{1}{(1 + i)^n}$$

where $\qquad i$ = rate of interest $\div$ 100
$\qquad\qquad\qquad n$ = number of years

In practice such calculations are usually done by taking the appropriate factor from a 'Present Value of £1' table and multiplying the future sum by this factor to obtain the discounted amount. The following is an extract from such a table:

Present Value of £1

Years	Interest		
	6%	9%	12%
5	0·7473	0·6499	0·5674
10	0·5584	0·4224	0·3220
20	0·3118	0·1784	0·1037
30	0·1741	0·0754	0·0334
60	0·0303	0·0057	0·0011

Thus if the cost of replacing a component in ten years' time is £100 and the rate of interest is assumed to be 9 per cent, the PV of the replacement is $0·4224 \times £100 = £42·24$.

Where regular annual sums are expended over a period of years the present value of the recurring sums may be obtained from the following formula.

$$\text{PV of £1 per annum} = \frac{(1 + i)^n - 1}{i(1 + i)^n}$$

Again it is more convenient to use factors from standard tables, e.g.:

Present Value of £1 per annum

Years	Interest		
	6%	9%	12%
5	4·212	3·890	3·605
10	7·360	6·418	5·650
20	11·470	9·129	7·469
30	13·765	10·274	8·055
60	16·161	11·048	8·324

Thus if the annual cleaning costs over the 60-year life of a building are £500 the present value, assuming a 9 per cent rate of interest, is $11 \cdot 048 \times £500 = £5524$. This table is also known for valuation purposes as 'Years Purchase – Single Rate' and can be used for finding the annual equivalent of an initial capital sum. For instance, if in the previous example the object had been to find the annual amounts which an investment of £5524 would yield over a period of 60 years at 9 per cent interest, the figure would be $£5524 \div 11 \cdot 048 = £500$. At the end of the period both the capital and interest would be expended unless an additional allowance had been made for a sinking fund.

Table 1.2. Discounted Cash Flow as Percentage of Initial Cost

| Year | % initial cost discounted at | | | | | |
| | 6% | | 9% | | 12% | |
		Cumula-tive		Cumula-tive		Cumula-tive
5	3·38	3·38	2·93	2·93	2·57	2·57
10	2·86	6·24	2·14	5·07	1·63	4·20
15	2·73	8·97	1·76	6·83	1·17	5·37
20	3·44	12·41	2·00	8·83	1·11	6·48
25	1·04	13·45	0·54	9·37	0·27	6·75
30	2·45	15·90	1·15	10·52	0·43	7·18
35	0·59	16·49	0·23	10·75	0·09	7·27
40	2·06	18·55	0·62	11·37	0·21	7·48
45	0·46	19·01	0·13	11·50	0·07	7·55
50	0·26	19·27	0·05	11·55	0·02	7·57
55	0·18	19·45	0·05	11·60	0·01	7·58
60	0·08	19·53	0·02	11·62	—	7·58

Applying discounting factors to the cash flows given in Table 1.1, the pattern of expenditure shown in Table 1.2 is obtained. The cumulative expenditure is shown in graphical form in Fig. 1.7 and it will be seen that as the interest rate increases the future maintenance costs appear less and less significant. This effect is even more pronounced when taxation is taken into account. As a general rule there is no tax relief for the initial capital cost of a building, but maintenance and running costs may be set against taxable profits. Thus, assuming a rate of 50 per cent for Corporation Tax, the true cost to

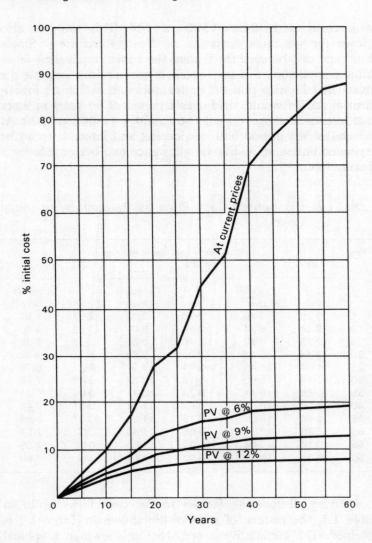

Fig. 1.7. Maintenance costs

the firm of maintenance is only 50 per cent of the actual cost. As opposed to this, maintenance and alterations are subject to Value Added Tax, but if the building owner is a 'taxable person' the tax is recoverable.

Maintenance profiles

The pattern of maintenance costs over the life of a building is known as the 'maintenance profile' and is usually presented in the form of a histogram. The example in Fig. 1.8 is based on the maintenance costs given in Table 1.1. The peaks of expenditure are attributable to major renewals and replacements which could be deferred by incurring higher running repair costs and perhaps some loss of amenity. Where the estate consists of a number of buildings some adjustment could be made to the frequency of renewals and cyclic work for the individual buildings so as to achieve a more uniform total workload from year to year. This would not only give a more acceptable cash flow but would be essential where it is the policy to do as much work as possible by direct labour.

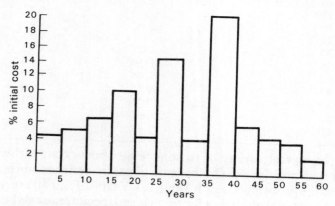

Fig. 1.8. Maintenance profile

However, it is somewhat unrealistic to relate predicted costs to particular years in the life of a building. Costs are made up of a large number of small items of uncertain timing together with a limited number of major items of renewal or replacement whose timing will depend, for the most part, on factors unconnected with the physical condition of the building. The uncertainties inherent in the timing of these major items can be reflected by assessing the probabilities of the work being necessary at different times. The probability scale runs from zero to unity, the bottom end of the scale representing absolute impossibility and the top end absolute certainty. In the context of maintenance few events are so clear-cut as this and in most cases the probability that the event will occur lies somewhere between these extremes. For instance, if there is a 50 : 50 chance that a particular component will have to be replaced in five years time, the probability would be expressed as 0·5, or in mathematical terms $p = 0·5$.

Applying probability theory to the renewal times for the felt roof covering, Fig. 1.6 indicates renewal in years 20 and 40

Table 1.3. Probable incidence of renewal times

Year	First renewal	Second renewal	Total probability of renewal
5	—	—	—
10	—	—	—
15	0·1	—	0·1
20	0·8	—	0·8
25	0·1	—	0·1
30	—	0·01	0·01
35	—	0·16	0·16
40	—	0·66	0·66
45	—	0·16	0·16
50	—	0·01	0·01
55	—	—	—
60	—	—	—

but for a great variety of reasons failure may occur earlier or later than anticipated. On the basis of past experience one might assess the probabilities as in Table 1.3. Years 20 and 40 are rated as the most probable times for the first and second renewals but, as the second renewal is dependent on the timing of the first, the probability that this will occur in the fortieth year is slightly less than the probability of the first renewal occurring in year 20. It has been assumed that as the life of the building is 60 years, renewal after the fiftieth year would not be contemplated.

The total maintenance cost for the roof will be made up of:

Renewal of the coverings
Patching of the coverings between renewals
Repairs to constructional members and eaves and rainwater goods.

as shown in Table 1.4.

Table 1.4. Maintenance cost of roof

Year	Renewal of coverings (assume cost of renewal including all labours £8 per m²) per 100 m²	£	Patching between renewals per 100 m² £	Repairs to structure, gutters, etc. per 100 m² £	Total per 100 m² £
5	—	—	20	20	40
10	—	—	40	20	60
15	0·1 × 800	80	60	20	160
20	0·8 × 800	640	20	20	680
25	0·1 × 800	80	40	20	140
30	0·01 × 800	8	60	20	88
35	0·16 × 800	128	60	20	208
40	0·66 × 800	528	40	20	588
45	0·16 × 800	128	60	20	208
50	0·01 × 800	8	80	20	108
55	—	—	80	20	100
60	—	—	80	20	100

The cost of patching has been related to the renewal periods but a constant sum has been assumed for repairs to structural members and rainwater gutters and pipes. This will only hold good if patching is carried out promptly so as to maintain the coverings in a waterproof condition. The maintenance profile of the roof is as shown in Fig. 1.9.

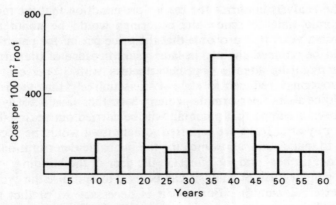

Fig. 1.9. Maintenance profile of roof

Similar maintenance profiles can be produced for other building elements to illustrate the timing and magnitude of the costs involved. Care should be exercised when applying costs and renewal cycles obtained from other buildings. Although the design and materials specification may be similar, the periodicity of repairing cycles may vary greatly according to climatic conditions and user activities. Even with the same building a change of use or an intensification of the existing use can have a profound effect on the maintenance profile. In particular, finishings and fittings may have a very much shorter life if subjected to more rigorous use conditions than those envisaged by the designer.

Optimising renewal cycles

The assessment of renewal cycles is based on knowledge of the rates of deterioration of similar buildings under similar conditions of exposure and use. Although the subjective element cannot be ruled out altogether it is possible to introduce a degree of objectivity into the assessment. One way in which this can be done is to prepare an Anticipated State Matrix for the element under consideration by assessing its probable condition at regular intervals throughout its life and estimating the cost of the remedial work necessary at those times. For instance, for the felt roof covering considered earlier the probable states and remedial costs may be as shown in Table 1.5.

The analysis confirms the earlier presumption that the most economic time to renew the coverings would be about the twentieth year. It is probable that the price per m² for patching would be reduced slightly in later years because of the larger areas requiring attention but nevertheless, with 30 per cent of the coverings reduced to state 4, it is unlikely that the roof would be able to perform its primary functions satisfactorily. It has been assumed that patching will be carried out reasonably promptly after leaks are reported: otherwise it would be necessary to make some allowance for the probable deterioration of the roof timbers and the effect on the user of the building.

A similar form of analysis can be used to optimise the cycles for external painting. For this it is necessary to predict the probable treatments required at intervals of one year up to, say,

Table 1.5. Renewal cycle of roof covering

Anticipated state	Remedial work	Cost per m²
1. No visible defects	Nil	—
2. Small isolated blisters	Nil	—
3. Large blisters and slight cracking causing minor localised leaks	Patching	£12
4. Extensive cracking and deterioration causing widespread leaks	Renew	£8

State	% of roof coverings in each state at year				
	5	10	15	20	25
1	80	60	30	15	10
2	19	25	35	25	20
3	1	10	25	30	20
4	—	5	10	30	50
Maintenance cost per m² (patching)	0·01 × 12 = £0·12	0·15 × 12 = £1·80	0·35 × 12 = £4·20	0·60 × 12 = £7·20	0·70 × 12 = £8·40*

* Cost of patching exceeds renewal cost.

ten years, and then identify the period which gives the lowest annual costs. The deterioration pattern will depend upon such factors as the quality of the paint, the thoroughness of the preparatory work, the severity of the exposure to atmospheric pollution and ultra-violet light and other matters peculiar to the particular building and the orientation of the elevation. For the purposes of this example it is assumed that the treatments required at the end of each period are as follows:

Year 1   Wash down
2   Wash down and one coat of paint
3   Wash down and two coats of paint
4   Wash down, burn off $2\frac{1}{2}$% and prime and two coats paint
5   Wash down, burn off 5% and prime and two coats paint
6   Wash down, burn off 10% and prime and two coats paint
7   Wash down, burn off 20%, renew $2\frac{1}{2}$% woodwork, prime and two coats paint

8 Wash down, burn off 40%, renew 5% woodwork, prime and two coats paint
9 Wash down, burn off 60%, renew 10% woodwork, prime and two coats paint
10 Burn off 100%, renew 20% woodwork, prime and two coats paint

The costs of the assumed treatments can then be calculated as shown in Table 1.6. On the basis of the assumptions made in this example the optimum period for external repainting would be six years. Washing down gives the same annual cost but even if this treatment were repeated each year it would be necessary after two or three years to apply one or two coats of paint to make good surface wear and defects due to movement of the substrate. Also, where appearance is of importance, it will be necessary to form a subjective assessment as to whether or not the state of the paintwork at the end of the economic period will be considered acceptable. The difference in cost

Table 1.6. Percentage of windows requiring treatment (based on 100 m² of softwood casement windows, measured overall)

Predicted treatments	Year 1	2	3	4	5	6	7	8	9	10
Wash down (£0.60 m²)	100	100	100	97½	95	90	80	60	40	—
One coat (£1.20 m²)		100								
Two coats (£2.40 m²)			100	100	100	100	100	100	100	100
Burn off (£4.50 m²)				2½	5	10	20	40	60	100
Prime & stop (£2.10 m²)				2½	5	10	20	40	60	100
Renew wood-work (£60 m²)							2½	5	10	20
Cost of treatment £	60	180	300	315	330	360	570	840	1260	2100
Cost per year £	60	90	100	79	66	60	81	105	140	210

between repainting at the calculated economic periods and at those judged necessary to maintain a satisfactory appearance will represent the value which the building owner or occupier attaches to appearance. Also the future costs could be discounted to allow for the time value of money.

Repair/replace decisions

Appraisal methods

There are several methods of analysing the economic consequences of alternative decisions relating to the repair or replacement of building elements and components. The more common methods are:

1. Payback period

The payback period is the length of time taken for the returns from an investment to equal the initial outlay. Where the proceeds are the same each year the payback period is obtained by dividing the initial outlay by the annual proceeds. Thus if roof insulation will cost £100 and the annual saving on fuel is estimated as £20 the payback period is five years. This can then be compared with the payback periods for other types of insulation, e.g. cavity-wall filling or double glazing to determine which type offers the best value. The method suffers from two main disadvantages:

(*a*) it does not take into account the time value of money,
(*b*) it ignores any benefits which may accrue after the end of the payback period.

It is, however, very simple and easy to understand.

2. Net Present Value (NPV)

This method is also known as the Present Worth method and involves discounting all future cash flows to a common base year. The formulae given earlier in this chapter for PV of £1 and PV of £1 per annum may be used for this purpose although it is probably easier to look up the factors in the standard tables. It should be noted that the analysis is particularly sensitive to the rate of interest used and, therefore, this must be chosen with care.

For example the total PV of repainting a building every five years over an anticipated life of 30 years (last repainting in year 25) will vary as follows for different rates of interest. For the purpose of this example the repainting cost has been estimated to be £200 and no allowance made for inflation.

PV of £1

Year	6% factor	9% factor	12% factor
5	0·7473	0·6499	0·5674
10	0·5584	0·4224	0·3220
15	0·4173	0·2745	0·1827
20	0·3118	0·1784	0·1037
25	0·2330	0·1160	0·0588
	2·2678	1·6412	1·2346
	× £200		
	= £454	= £328	= £247

It will be noted that the higher the rate of interest the lower the PV of future costs and the less important maintenance costs appear to be in comparison with the initial costs of construction. Also, if the above analysis were being used to determine how much it would be worthwhile spending on finishes which would not require periodic repainting the decision would obviously depend upon the rate of interest used.

The method may also be used to produce a modifed form of the payback method which takes into account the time value of money. Thus the example given in which the expenditure of £100 on insulation results in a £20 saving on fuel costs per annum could be analysed as follows:

Year	Saving (not adjusted for inflation)	PV of £1 factor for 9%	PV of un-adjusted saving	Saving adjusted for 10% inflation per annum	PV of adjusted saving
	£		£	£	£
1	20	0·917	18·34	20	18·34
2	20	0·842	16·84	22	18·52
3	20	0·772	15·44	24·2	18·68
4	20	0·708	14·16	26·6	18·83
5	20	0·650	13·00	29·3	19·05
6	20	0·596	11·92		
7	20	0·547	10·94		93·42
			100·64		About 5½ years
			7 years		

Thus, if no account is taken of inflation, the payback period when considering the present values of the future savings is seven years. If a rate of inflation of 10 per cent is assumed the payback period is reduced to about five and a half years. A larger increase in the price of fuel would reduce the period still further and strengthen the case for providing the insulation.

3. Annual Equivalent (AE)

In this method the cash flows throughout the life of the asset are converted into an equivalent uniform annual cost. It will give the same ranking as the Present Value method but the presentation in the form of annual outgoings (or receipts) is probably more meaningful to the building owner. Where the cash flows are irregular it is easier to convert them all to their present value initially (except those which are regular annual payments and do not require converting) and then calculate the annual equivalent of the total present value. This is done by multiplying the total present value by the reciprocal of the PV of £1 per annum factor. This factor is known as the Uniform Series that £1 will buy and is given by the following formula.

$$\text{Uniform Series that £1 will buy} = \frac{i(1 + i)^n}{(1 + i)^n - 1}$$

Again the rate of interest can have a significant effect on the ranking of alternative courses of action. Also difficulties can arise when comparing alternatives which have different lives. In such cases it is usual to base the analysis on a period of time which is the lowest common multiplier of the lives of the alternatives. Thus if a comparison is being made between a traditional brick building with an assumed life of 60 years and a timber-framed building with lightweight claddings and an assumed life of 30 years the analysis would be based on a 60-year period. This would necessitate including for the renewal of the shorter-lived building at the end of 30 years. Alternatively, if the analysis is based on the 30-year period an allowance should be made for the value of the traditional building at the end of that period otherwise the comparison is not being made on a common basis.

In some cases payments increase by a regular amount each

year. The annual equivalent of such increases can be found using the following formula.

$$\text{Uniform gradient series} = \frac{g}{i} - \frac{ng}{i}\left(\frac{i}{(1 + i)^n - 1}\right)$$

For example, assuming the maintenance costs for a building are £2000 in year 1 and increase by £50 per annum throughout its 60-year life, the annual equivalent of the increase at 10 per cent interest will equal:

$$\frac{50}{0\cdot1} - \frac{60 \times 50}{0\cdot1}\left(\frac{0\cdot1}{(1\cdot1)^{60} - 1}\right) = £490$$

This added to the regular component of £2000 gives a total annual equivalent cost of £2490.

4. Yield method

This is also referred to as the Internal Rate of Return method or in some textbooks as Discounted Cash Flow (DCF) although strictly speaking DCF would also include the PV method. The main difference between this method and the previous two methods is that in the Present Value and Annual Equivalent methods a fixed interest rate has to be assumed whereas the object of the Yield method is to find the interest rate which will equate the total present value of all the future cash flows, both negative and positive, to the initial outlay.

For example, it has been estimated that the provision of certain mechanical handling equipment in a warehouse would reduce labour costs by £6000 per annum. The cost of installing the equipment is £20 000 and will require additional running costs of £1000 per annum. What is the

Year	Receipts £	Payments £	Net Cash Flow £
0	—	−20 000	−20 000
1	+6000	−1000	+5000
2	+6000	−1000	+5000
3	+6000	−1000	+5000
4	+6000	−1000	+5000
5	+6000	−1000	+5000
	+30 000	−25 000	+5000

rate of return from the investment assuming that the warehouse will only be used for a further five years? The first thing to do is to find the net cash flows over the period as shown in the preceding table. Then it is necessary to find the interest rate which will discount the future cash flows to give a total PV which is equal to the initial outlay of £20 000. This is found by trial and error although the following formula will give an approximate value for short periods:

$$\text{Interest rate} = \frac{\text{Total proceeds} - \text{Initial cost}}{\text{Period in years} \times \text{initial cost}} \times 100$$

(multiply by 2 for longer periods over 10 years)

The formula gives an approximate rate of 5 per cent and this is used as a starter. As the total PV so obtained is greater than the initial outlay, i.e. £21 648, the true interest rate must be higher than 5 per cent. Therefore, a rate of 9 per cent is tried and this time the total PV is less than the initial outlay, indicating that the required interest rate must lie between 5 and 9 per cent. The actual rate is found by interpolation as follows:

$$\text{Yield (\%)} = 5 + \left[\frac{21\ 648 - 20\ 000}{21\ 648 - 19\ 449} \right] \times (9 - 5)$$

$$= 5 + \frac{1648 \times 4}{2199} = 8\cdot06\%$$

Year	Net cash flow NCF	PV of £1 factor 5%	PV of NCF at 5%	PV of £1 factor 9%	PV of NCF at 9%
	£		£		£
0	−20 000	1·0000	−20 000	1·0000	−20 000
1	+5000	0·9524	4762	0·9174	4587
2	+5000	0·9070	4535	0·8417	4209
3	+5000	0·8638	4319	0·7722	3861
4	+5000	0·8227	4114	0·7084	3542
5	+5000	0·7835	3918	0·6500	3250
	+25 000		+21 648		+19 449

The calculation can be checked by discounting the cash flows using an interest rate of 8 per cent when the total PV should equal £20 000 (approx.). Whether or not the investment is considered worthwhile will depend upon the rate of return which can be obtained from an alternative use of the capital.

5. Break-even analysis

This method is concerned with making comparisons between two alternatives where the cost of each alternative is affected by a single common variable. The comparison may be made either mathematically or graphically and involves finding the point at which the value of the two alternatives is the same.

For example, a choice is to be made between two components A and B in which A has a lower first cost but higher maintenance costs than B. This can be represented as in Fig. 1.10 which shows that if the period of use is less than 20 years A is more economic whilst if greater than 20 years B is to be preferred.

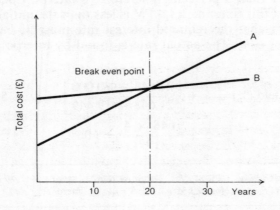

Fig. 1.10

6. Decision trees

The decision tree attempts to account for future uncertainties by evaluating all possible outcomes using subjective probabilities and expected values. The expected value of an event is its probability of occurrence multiplied by its conditional value. The conditional value being the loss or gain if the event in question actually occurs. Thus if the consequential losses flowing from the breakdown of a machine are estimated to be £1000 and the probability of a breakdown occurring is

0·1, the expected value is £100. This gives some measure of the amount which it is worthwhile spending in order to avoid such a breakdown.

7. Simulation

This involves designing a model, usually utilising a computer, which closely resembles the operation being studied. Many variables are built into the model but no attempt made to estimate their precise values or frequencies of occurrence. Then the variables are systematically altered within their possible ranges in order to determine the effect of the different values on the working of the model and to identify the optimum solution.

Replacement

Generally replacement of a component will be considered when either

(*a*) it is functionally or aesthetically unsatisfactory, or
(*b*) the repair and running costs are excessive.

The choice is then between replacement with an identical component or with one of a different type which has a better performance. In such cases comparative costs-in-use analyses can be prepared to assist in deciding which alternative offers the best value for money. Often, however, a component is repairable and the decision as to whether to repair or replace will depend upon the repair costs and lives of the alternative courses of action. If it is assumed that the life of the repaired component is shorter than that of the replacement then it is quite simple to calculate the maximum economic amount to spend on the repair. For instance if the repair life is five years and a replacement would last ten years it would not be economic to spend more than 57 per cent of the cost of replacement on the repair, assuming an interest rate of 6 per cent.

This is calculated as follows:

Assume £x = repair cost, and   Interest rate = 6%
          £y = replacement cost

£x |____|____|____|____|____|

years →   5

$$\text{Annual equivalent} = \frac{x}{4 \cdot 212*}$$

£y ————+————+————+————+————+————+————+————+————+————

years →   10

$$\text{Annual equivalent} = \frac{y}{7 \cdot 360*}$$

* from PV of £1 per annum (Years Purchase) Table

For x and y to give equal value

$$\frac{x}{4 \cdot 212} = \frac{y}{7 \cdot 360} \qquad x = 0 \cdot 572y$$

Therefore, it is uneconomic if x, the repair cost, is more than 57·2 per cent of y, the replacement cost.

For ease of reference this can be set down in tabular form as in Table 1.7. As will be seen, the higher the rate of interest the more worthwhile it is to take the shorter-term view and spend a higher proportion on repair, even though the life of the repaired component is shorter than a replacement.

However, where the replacement is of a more advanced design than the original any ensuing benefits must be taken into

Table 1.7. Maximum repair cost as percentage of replacement cost

Repair life (years)	Replacement life (years)				
	10	15	20	25	30
Interest 6%					
1	12·8	9·7	8·2	7·4	6·9
2	25·0	18·9	16·0	14·3	13·3
3	36·3	27·5	23·3	20·9	19·4
4	47·1	35·7	30·2	27·1	25·2
5	57·2	43·4	36·7	32·9	30·6
Interest 9%					
1	14·3	11·4	10·0	9·3	8·9
2	27·4	21·8	19·3	17·9	17·1
3	39·4	31·4	27·7	25·8	24·6
4	50·5	40·2	35·5	33·0	31·5
5	60·6	48·3	42·6	39·6	37·9

account. This is particularly so where the replacement offers a saving in running costs. For example, if it is to be decided whether to repair an existing boiler at a cost of £200 or replace it with a more efficient type at a cost of £500 the 'total costs' of the two alternatives must be calculated. If it is assumed that the repaired boiler will have a life of ten years and consume £250 of fuel each year whereas the replacement will have a life of 15 years and consume only £200 of fuel each year, the total costs will be as follows:

Repair

Annual equivalent of £200 over 10 years at 6% interest

$$= \frac{200}{7 \cdot 360} \qquad 27$$

$$\text{Fuel costs} \qquad \underline{250}$$

$$\text{Total annual cost} \quad £277$$

Replacement

Annual equivalent of £500 over 15 years at 6% interest

$$= \frac{500}{9 \cdot 712} \qquad 51$$

$$\text{Fuel costs} \qquad \underline{200}$$

$$\text{Total annual cost} \quad £251$$

A replacement boiler would therefore represent the better value. This method of analysis may also be used when considering changing a boiler for one which uses a different type of fuel, although a forecast would have to be made of the future price levels of the fuels in order to identify the most advantageous long-term solution.

Similarly, the high cost of electricity may make it worthwhile to replace tungsten lamps with fluorescent tubes. The amount which it is worthwhile spending on effecting the change can be determined by calculating the savings which would be achieved over an appropriate period. For example, assume that a space is lit at present by means of one hundred 150 W tungsten lamps and it is required to find out how much it would be worthwhile spending on replacing them with a similar number of 40 W fluorescent tubes. It is estimated that the lighting will be used for about 1000 hours per year and that the comparative data are as follows:

	Life Hours	Bulk replacement Period	Replacement costs per lamp/tube £	Electricity consumption W
Tungsten lamps	1000	Annually	0·50	150
Fluorescent tubes	7000	Every 7 yrs	2·50	50 (extra 10 W consumed in control gear)

Analyses based on assumed life for fitting of 14 years and a rate of interest of 9%

Tungsten lamps
Annual cost

Bulk replacement	100 lamps @ £0·50 each	50·00
Electricity	$\dfrac{100 \times 150 \times 1000}{1000}$	
	= 15 000 units @ 4p per unit	600·00
		650·00
Factor for PV of £1 p.a. for 14 years @ 9%		× 7·786
	Total PV =	£5061

Fluorescent tubes
Annual cost

Electricity	$\dfrac{100 \times 50 \times 1000}{1000}$	
	= 5000 units @ 4p per unit	200·00
Factor for PV of £1 p.a. for 14 years @ 9%		× 7·786
		£1557
Bulk replacement after 7 years		
100 tubes @ £2·50 each	£250	
Factor for PV of £1 in 7 years @ 9%	× 0·5470	137
	Total PV =	£1694

Saving over 14 years = £5061 − £1694 = £3367
(Cleaning costs assumed to be the same for both types of fittings)

Thus it would be worthwhile spending up to about £33 per point on the change. If the lighting is used for more than the estimated 1000 hours per year or if there is an increase in the price of electricity the change becomes even more worthwhile.

Other possibilities that could be considered in order to reduce electricity consumption on lighting range from the simple one of ensuring that lights are switched off when not required to providing separate switching for different areas with separate task and ambient lighting systems. Also, one well-known departmental store found that substantial savings could be made by reducing the level of illumination in the sales areas from 1000 lux to 600 lux without any detrimental effect on sales.

Alterations

Decisions relating to alterations are based on similar considerations to those which apply to repair or replace decisions. In both cases the initial costs of the changes in terms of their annual equivalent plus the future running costs should be less than the existing running costs.

The reasons for alterations include:

(*a*) Increasing the value of the asset.
(*b*) More convenient or effective use of the building.
(*c*) Reduction in overall costs (maintenance, energy, wages, insurance, etc.).
(*d*) Better appearance.
(*e*) Compliance with statutory requirements.

Significant savings can often be made by amending existing layouts within the structural limitations of a building. Studies of the movements of workpeople using such devices as string diagrams may show that a rearrangement of the work spaces and equipment would cut down the frequency and length of journeys between essential points. Assuming that the total annual costs are broken down as follows:

	%
Buildings and equipment (amortised)	8
Maintenance, heating, lighting, cleaning, etc.	22
Staff salaries	70
	100

If movement accounts for one-third of staff time the cost would be about 23 per cent the total annual cost. A saving of 20 per cent in the amount of movement would thus give about a 5 per cent reduction in total costs – a saving which would be difficult to achieve in any other way.

For example, it has been estimated that changes in the layout of an industrial kitchen costing £10 000 would result in savings in labour and materials of £5000 per annum. It is required to determine the period of time over which the change must be satisfactory in order to justify the expenditure. The period for which the alteration should be satisfactory is given by the equation:

Difference in running and operating costs

$$= \frac{\text{Cost of alterations}}{\text{PV of £1 p.a. (or Years Purchase)}}$$

In the example

$$5000 = \frac{10\,000}{\text{PV of £1 p.a. factor}}$$

$$\text{PV of £1 p.a. factor} = \frac{10\,000}{5000} = 2 \cdot 0$$

It is then necessary to consult the standard tables for the PV of £1 p.a. and find the period which has a factor of 2·0 at the required rate of interest.

Thus at 6 per cent interest the factors are as follows:

for 2 years the factor is 1·833, and
for 3 years the factor is 2·673.

The period for a factor of 2·0 is between 2 and 3 years and is found by interpolation to be 2·2 years.

At 12 per cent interest the factors are as follows:

for 2 years the factor is 1·690, and
for 3 years the factor is 2·402.

Again the period for a factor of 2·0 is between 2 and 3 years and is found by interpolation to be 2·4 years.

Other methods of appraisal could be used for this sort of problem. For instance, the payback method would give a period of two years which might be sufficiently accurate for the purpose. Also, if the period over which the change will be effective is known, the yield from the investment could be calculated for comparison with other investment opportunities.

Conversions

The existing stock of buildings represents a considerable national asset and clearly should be used as effectively as possible. The public discussion has usually centred around housing and the undesirable effects of breaking up established communities by the comprehensive redevelopment of slum areas. The argument is that rehabilitation would not only be cheaper but would avoid some of the social problems that beset new estates. However, the working environment is just as important and the same sorts of problems arise in the case of old commercial and industrial buildings. In many cases by rearranging the internal spaces and providing new facilities they can be given a new lease of life by making them suitable for some new use. This process is called by a wide variety of names – adaptation, conversion, refurbishing, retrofitting, rehabilitation, renovation, modernisation, etc. – representing the different approaches to the problem of making old buildings fit for new uses.

Basically the decision as to whether to adapt an old building or demolish it and erect a new building will be based on the same sorts of economic analyses as outlined earlier in this chapter unless the building is listed or in a conservation area or if there are other non-economic reasons for its retention. Naturally, consideration will have to be given to the physical condition of the buildings and its probable life expectancy and whether or not the form of construction will permit the necessary structural changes to be made. Also the size of the spaces within the building and the ease with which they can be changed would be important factors. Generally the old services will be inadequate and will have to be completely renewed. However, refurbishing is usually a quicker process than complete rebuilding and may offer other advantages, e.g. the old building might enjoy a higher plot ratio than would be permitted by the planning regulations for a new building on the same site or it might be thought desirable to preserve the character of the existing environment.

Energy management

The maintenance manager is likely to be involved in decisions concerning alternative ways of energy conservation. This will usually start with the preparation of an energy budget or audit giving the energy consumption of the building from all sources for internal environmental control stated in units of energy per unit of floor area. Comparisons can then be made of the energy usage of different areas and buildings to identify high cost areas.

For inter-firm comparisons some assistance may be obtained from the Energy Cost Analyses distributed to subscribers by the Building Maintenance Cost Information Service. The energy budget will only indicate the amount of energy being used and not whether it is being used effectively. It is necessary, therefore, to carry out an investigation of the way in which energy is being used and to determine the measures that should be introduced to avoid waste. The measures include:

1. Good housekeeping. This involves making the people who use the building aware of the need to conserve fuel by such simple measures as switching off the lights when they leave a room unoccupied. Perhaps it involves persuading cleaning staff that it is not really necessary to have all the lights on throughout the entire building for the whole of the period that they are working in the building at night.
2. Reviewing standards. An examination should be made of the standards of heating, lighting, ventilation, humidity, etc. to make sure that they are not unnecessarily high.

3. Modifications to existing systems. These may be quite minor modifications such as excluding draughts from windows and external doors or providing additional insulation. Others could be fitting thermostats to radiators or photo-cells to control lighting installations.
4. Provision of new equipment. This would cover more ambitious schemes such as heat exchangers to recover heat from exhaust air or waste water, computerised controls to optimise the firing of boilers, and use of solar energy.

The financial consequences of the above measures can be determined using the appraisal methods described earlier. It should be mentioned that much more could be done at the initial design stage to provide energy efficient buildings by giving more thought to the basic shape, form and orientation of the building and by a more integrated approach to the design of the fabric and the services.

References

[1] HMSO (1981) *English House Condition Survey 1981:* HMSO.
[2] National Institute of Economic and Social Research (1970) *Urban Development in Britain, 1964–2004.* Cambridge University Press.
[3] Stone, P. A. (Feb. 1970) 'Economic Realities.' *Official Architecture and Planning* 131–4.
[4] Needleman, L. (1965) *The Economics of Housing.* London: Staples Press.
[5] Stone, P. A. (1980) *Building Design Evaluation – Cost-in-Use.* London: Spon.
[6] Noblé, V. (1973) 'Is life forecasting possible and/or worthwhile?' London: National Building Maintenance Conference.
[7] Switzer, J. F. Q. (1963) 'The life of buildings in an expanding economy.' RICS Conference.
[8] HMSO (1972) *Report of the Committee on Building Maintenance.* Department of the Environment: HMSO.
[9] HMSO (1970) *Report of the Committee on Hospital Building Maintenance.* (Chairman, David Woodbine Parish.) Department of Health and Social Security: HMSO.

10 White, D. J. (1969) 'Management science for maintenance.' London: Conference on Building Maintenance.
11 Bushell, R. J. (1979/80) 'Preventing the problem – a new look at building planned preventive maintenance.' Institute of Building Information Service **11**.
12 Flanagan, R., and Norman, G. (1983) *Life Cycle Costing for Construction*. London: RICS.

CHAPTER TWO

Maintenance standards

General considerations

The identification of appropriate building standards is a key factor in determining the maintenance workload. The concept of an 'acceptable standard' is mentioned in a number of the definitions previously discussed, but this does not imply that there is any absolute standard which would be satisfactory in all cases. While some requirements may have universal applicability, e.g. structural stability, others would have to be assessed on economic and social grounds according to the overall policies of the owner or occupier. Robertson[1] names some of the factors which would have to be taken into account as 'political, legal and financial considerations, forecasts of future activity, market conditions, taxation policy and labour conditions'. Clearly the different requirements of different organisations will make different maintenance policies inescapable.

Expression of standard

It is attractive to think in terms of a range of acceptability, e.g. Kemp[2] suggests the fixing of upper and lower criteria as illustrated in Fig. 2.1. The higher limit is set by the cost of achieving it and the lower limit by the increasing probability of failure involving not only enhanced repair costs but also consequential losses where the normal user of the building is interfered with. The model is apparently related to functional performance and assumes a uniform rate of deterioration which will eventually result in failure. However, this is not an accurate representation of the behaviour of many building elements. Certain elements maintain a constant condition over the life of the building, while others are subject to sudden unpredictable

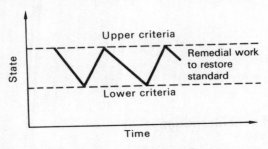

Fig. 2.1. The passage of time related to maintenance criteria

failure. In some cases there is only a single criterion, e.g. whether or not the roof leaks, while in others there is no precise lower limit, e.g. decorations do not 'fail' in such a way as to impede user activities but must be judged on the basis of visual acceptance. Also, over the long life of a building, users will tend to come to demand higher standards and therefore repairs and renewals will inevitably contain some element of improvement.

The model shown in Fig. 2.2 accords more closely with reality, although it needs adapting to the characteristics of the particular element.

Having considered the particular requirements for each individual element, the standard may be expressed in the following ways:

1. Physical terms relating to:

   (*a*) The condition of the element specifying the magnitude of the defects which call for remedial action. Wherever possible, the extent of the defect should be capable of direct measurement to ensure uniformity of interpretation where more than one person is responsible for inspections.

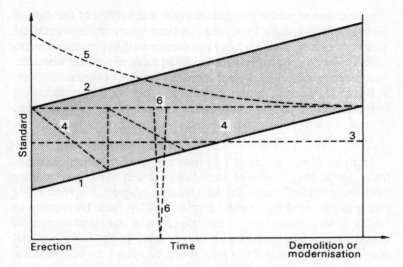

Fig. 2.2. Maintenance requirements related to standards over a period of time
(Source: ECE. Proceedings of Seminar on Management, Maintenance
and Modernisation of Housing, Warsaw. United Nations, 1969)

    1. Lowest acceptable standard during time of use. The upward curve of
the line symbolises change in the level of requirements

    2. Optimal standard during time of use; the area between lines 1 and 2
is the accepted standard area

    3. Element with immutable quality, i.e. horizontal line

    4. Rapid wear leading to maintenance during time of use, e.g. wall-
paper

    5. Slow, undramatic change during time of use. Can be compensated
by providing a higher standard from the beginning

    6. Dramatic failure calling for immediate action, e.g. leakage in water
or sanitary installations

(*b*) The performance of elements or environmental systems.
Where performance specifications have been used for the
initial design, these could be adopted for this purpose.

This system is called 'condition controlled maintenance' and it
presupposes that there will be inspections at appropriate intervals in order to determine by visual means or measure whether
or not the condition of the elements or their performance has
deteriorated below that laid down.

2. Times at which repairs and replacements are to be made.
This method is sometimes referred to as 'frequency based

maintenance' and it requires certain knowledge of the rate of deterioration and of the point in time when either functional failure is imminent or the appearance will become unacceptable. Clearly a proper balance must be achieved between the frequency and the risk and consequences of failure.

3. Financial criteria which may take the form of a variable sum related to the cost of some primary activity or replacement value or a fixed sum based on historic costs or an analysis of anticipated benefits.

In most cases, all three methods are used, although there is little doubt that financial expediency often takes precedence over the physical needs of the building. A possible reason for this was revealed in a study carried out by Bath University in which it was found that in the larger firms the initial technical assessment was carried out at a low level in the firm, but that the ultimate decision as to how much to spend on maintenance was taken at a higher level and was in many cases little more than a trimming exercise. The budget was generally the amount expended the previous year with, perhaps, some adjustments for increased costs. While not denying the need for financial constraints, there should be greater awareness of the ultimate cost of under-maintenance. Thus instead of thinking in terms of the current availability of funds the total costs over a substantial period, say ten years, of alternative maintenance policies should be compared.

Application of criteria

Two reported cases serve to show how the amount of maintenance is usually determined. The first concerns the maintenance of retail outlets, for which two sets of figures are produced:

1. The amount which the surveyor would like to spend on the property to put it into first-class structural and decorative repair.
2. The amount which it is considered economic to spend on each property in view of its profitability.

If there is gross disparity between the amount based on the physical condition of the building and that based on financial considerations, the future of the property is considered. The main disadvantage would seem to be that the least profitable outlets are likely to be the most dilapidated and yet progressively less and less money will be spent on their repair. Also, the surveyor bases the extent of the work on a subjective notion as to what constitutes a first-class condition. In many cases economic considerations will dictate a lower standard and therefore conflict between the two amounts is inevitable.

The second case relates to the introduction of a planned maintenance system for local authority housing. Again, this is in two stages:

1. The houses are inspected by qualified inspectors who describe the defects in each property and then the work is costed to provide the inspector's 'ideal' budget, i.e. the amount necessary to put the property in first-class order.
2. A financial limit is determined for each property on an age/cost basis, and, if necessary, the ideal budget trimmed by omitting some of the less essential items. The age/cost amount is determined by dividing the total annual appropriation by the total age of all the houses in the area and then multiplying this by the age of the individual house under consideration.

This system assumes that there is a direct relationship between age and maintenance costs and that the total sum allocated is the correct amount. Also, there should be a system of assigning priorities to the various items so that the less important ones can be clearly identified.

These methods correspond broadly to the recommendation by Ray[3] that the problem should be viewed in two parts:

1. The establishment of authorised expenditure limits related to anticipated returns.
2. The effective and efficient use of resources involving a technical assessment of the condition of the building carried out within a priority system which will show not only what needs to be done, but also the consequences of deferring work.

Attitude of maintenance initiators

In practice much will depend upon the attitude and status of the person responsible for initiating maintenance work. The primary initiators of maintenance are the owner and/or occupier, although other interested parties, e.g. factory inspectors, insurance companies, employees and their trade unions, members of the public, etc., may exert either a direct or an indirect influence on the amount of work undertaken.

Owner

The attitudes associated with the owner function will be broadly similar whether the owner is in occupation or leasing the building. The primary aim is to preserve the value of the asset so as to ensure a long-term trouble-free investment capable of providing a continuous and satisfactory return. The object is to achieve this with the minimum expenditure. Ideally, advance provision should be made for future repairs which should be preplanned so as to achieve an even cash flow over the years.

However, if the building is viewed purely as an investment then clearly the return should be no less than that obtained from alternative investment opportunities. In the absence of current or future benefits, there is no economic incentive for the owner to do more than is necessary to comply with statutory and contractual obligations. Also, the owner's long term interest in the property leads him to place greater emphasis on structural repairs rather than the internal decorative condition which has little direct effect on the rate of deterioration of the property. However, there may be indirect consequences in that the degraded internal appearance may result in less careful treatment by the occupier.

Owners Interest may not be long term.

Occupier

The attitudes of the occupier are distinguished here from those of the owner although both would be present in the case of an owner/occupier. Clearly the maintenance policy must be related and subservient to the fundamental aims and objectives of the individual or organisation which occupies the building. In order to achieve its objectives an organisation must initiate and sustain certain activities or modes of behaviour which require the provision and maintenance of a compatible spatial and physical environment. The maintenance needs of the organisation may therefore be stated at four levels of specificity:

1. The effect of maintenance standards on the objectives of the organisation. The main aim of most organisations is survival which, in a commercial context, is dependent upon making a profit. However, there is a growing need for criteria which go beyond simple economic analysis and take into account the effectiveness of maintenance in relation to corporate objectives broader than that of profit maximisation or which reflect more sophisticated ways of ultimately achieving maximum profit. The standard must therefore reflect the social attitudes towards the environment in which people live and work and is only partly determinable on a rate-of-return basis.
2. The effect on the activities necessary to achieve the desired economic or social objectives. This is relatively straightforward where the purpose of the building is to house manufacturing processes which require rigorously controlled conditions of temperature, humidity, dust particles, etc., in that deviations from the 'norm' may have measurable effects on output. Human beings are much more adaptable and their reactions to changes in the environment not entirely predictable except in extreme conditions. An understanding of the physiological and psychological factors underlying human sensation and perception is necessary in order to predict the effects of marginal changes in environmental conditions on user activities.
3. The effect on the internal environment in which the user activities take place. This involves consideration of the total environment produced by the building, the people and process machinery which it houses and the external environment. The cost of maintaining the environmental conditions at the required level may be significantly affected by deficiencies in the state of the building fabric. For example, the

greater heat loss through damp insulation caused by a roof leak must be made good by the input of heat with correspondingly higher fuel costs.
4. The effect on the physical condition of the building elements and materials of construction. In particular, the long-term effects of defects on the element affected and on adjoining elements should be considered.

At present the initial technical assessment is carried out at level 4 and expenditure limits at level 1, but little attempt is made to trace the effects of disrepair through the intervening levels. The result is the two-part approach described earlier and the apparent conflict between the maintenance organisation and upper management.

Specific performance standards for analysis at level 3 exist in codes of practice, regulations and research recommendations, although in many cases these appear to be based more on preconception and intuition than on a scientific study of level 2 criteria. The weak link in the chain is undoubtedly lack of knowledge of the way in which the environment affects the behaviour of people. Without this knowledge, it is difficult to relate maintenance standards precisely to the degree of fulfilment of objectives.

Division of responsibility

A landlord normally seeks to preserve the condition of his property by the insertion of appropriate clauses in the lease. However, in some cases the position is regulated by statute. Thus, Section 8 of the Landlord and Tenant Act, 1985, provides that in the case of certain small dwellings let for a period less than three years there shall be an implied term in any contract of letting that the house is at the commencement of the tenancy 'fit for habitation' and shall remain so during the tenancy. The criteria for determining fitness for habitation are as stated in Section 4 of the Housing Act, 1957 (now re-enacted in Section 604 of the Housing Act, 1985).

Section 11 of the Landlord and Tenant Act, 1985, provides that in any lease for less than seven years the landlord shall:

1. keep in repair the structure and exterior of the dwelling (including drains, gutters and external pipes),

2. keep in repair and proper working order installations for the supply of water, gas and electricity and for sanitation (including basins, sinks, baths and sanitary conveniences but not other appliances which make use of the supply of water, gas and electricity),
3. keep in repair and proper working order the installations for space heating and heating water.

The standard required is a variable one based on the age, character and prospective life of the house and the locality in which it is situated. The condition of the other houses in the area of a similar type is probably the main criterion used when determining an appropriate standard.

At the moment, there is no similar code imposing specific responsibilities for repair and maintenance upon landlords and tenants of other types of property. The usual basic terms in a modern commercial lease require the tenant to keep the building in good and substantial repair and to redecorate both the interior and exterior at certain intervals, usually three to five years for the exterior and five to seven years for the interior. It should be noted that the expression 'keep in repair' also means 'put in repair' and that it is a wise precaution to see that a schedule of condition is attached to the lease in order to limit the liability both during and at the end of the lease. Generally such a repairing covenant does not impose a liability to provide something which was not there before. Thus, if there is no damp-proof course, the lessee is not obliged to provide one although he would be responsible for remedying defects which arise because of this omission. A more onerous provision is that the lessee should be liable for 'inherent defects' which would extend to making good design defects. Formerly the expression 'fair wear and tear excepted' was commonly used in leases for short terms. This has been legally interpreted as defects caused by the normal action of the elements or the normal human use of the premises. However, the lessee is bound to do such repairs as may be required to prevent the consequences flowing originally from wear and tear from producing others which wear and tear would not directly produce.

Where buildings are let in multiple occupation, it is normal to require tenants to pay a service charge to cover among other things the maintenance of the structure, the exterior and common parts of the interior for which the landlord is responsible. Such a charge should be subject to review at fairly frequent intervals so that adjustments can be made for increases in the cost of carrying out the necessary work. In the case of flats the Landlord and Tenant Act, 1985, provides that the costs must be reasonably incurred and that the services or work must be of a reasonable standard. The Act further protects the interests of the tenant by giving him a right to require the landlord to supply a summary of the costs incurred and to inspect the accounts and receipts on which the summary is based. Where the tenant has a liability as part of the service charge to pay for repairs he has the additional right to be consulted before major works are carried out by the landlord, who is required to obtain at least two estimates for the work.

An important factor is that the lease provides a means whereby the landlord can impose a system of planned maintenance upon the tenant to ensure that the condition of the property does not deteriorate. The provisions can be enforced by reserving a right of entry to inspect the property during the period of the lease and by requiring the premises to be delivered up at the end of the lease in the same condition that they were in at the beginning of the tenancy. In order to enforce the requirements of the lease the landlord may serve on the tenant a Schedule of Dilapidations. There are two types of schedules:

1. Interim Schedule of Dilapidations served during the currency of a lease together with a notice to repair. This is a list of all the items of repair necessary under the terms of the lease and is usually served with the intention of enforcing a right of re-entry if the tenant fails to carry out the work. Some relief is provided in the case of residential property under Part 2 of the Landlord and Tenant Act, 1954, where there are more than three years of the lease remaining unexpired. In this case the tenant may apply to the court for relief and the landlord must show that in the absence of compliance he would suffer an immediate damage to his reversion.
2. An End of Lease Schedule which forms the basis of a money claim against the tenant in respect of his breach of

covenant to keep and leave in repair. This is similar to the interim schedule but the items of repair are priced out to form the basis of a lump sum claim for compensation in lieu of the tenant carrying out the repairs.

Determinants of maintenance standards

The factors which influence decisions to incur maintenance expenditure are complex and in some cases conflicting. However, the factors should be made explicit in a planned maintenance policy spanning a number of years. The factors include the satisfaction of user requirements, value considerations and statutory constraints.

Basic user requirements

A stable shelter from the weather
The earliest buildings were designed to do little more than protect the occupants from the vagaries of the climate. With the increasing complexity of buildings, the walls and roof have assumed a variety of other interdependent functions. However, freedom from collapse remains an absolute requirement for all buildings. Some movement is inevitable and it is necessary to determine whether the amount of movement and the rate at which it takes place is indicative of imminent failure of a structural member. Similarly, some degree of damp penetration may be tolerated in certain situations. In such cases, the appropriate standard would be determined partly by visual acceptance and partly by assessing the effect of dampness on the functional efficiency of the building envelope.

Suitable internal environmental conditions
These are usually provided partly by natural means and partly by artificial means. The external envelope may be regarded as a filter which can be adjusted to allow varying amounts of light and air into the building while excluding undesirable features, such as noise and excessive heat. Where the input from natural sources is inadequate, the balance must be supplied artificially. There is thus a continuous interplay between the fabric of the

building and the services in regulating the internal environment.

Distinguishing levels of acceptability

Stevens[4] draws attention to the difficulty of making rational decisions about the appropriate level of maintenance expenditure. He proposes that assessments of building condition should be made on a numerical basis and divided into five classes ranging from very good to dangerous. Similarly, he suggests that the desirable level of maintenance should be expressed in five classes ranging from very high to very low according to the use of the particular area of the building. A comparison of the actual condition of the part of the building under consideration and the desirable level of maintenance for that part will indicate the priority to be accorded to the work.

Limiting this to the habitable or working areas of a building it is possible to distinguish three broad levels of intent:

Lowest level. The immediate environment should not be directly harmful or uncomfortable.

Middle level. The environment should be such as to minimise effort at work or maximise output.

Highest level. The environment should promote not only the actual wellbeing of the users of the building, but also their sensation of wellbeing.

Lowest level

Achieving the lowest level was the task of the nineteenth-century reformers who were in the main medical men rather than architects. The bringing together of large numbers of people and the pollution of the atmosphere by industry gave rise to environmental problems of the most compelling urgency. As early as the 1860s the difference in health of those working in controlled environments, even if crudely controlled, and those working in relatively uncontrolled ones was a matter of public record. The public health concepts developed during this period form the basis of much of our present-day legislative controls.

Middle level

The early reformers were thinking purely in terms of health and safety, but clearly if conditions were so inadequate that they increased the sickness and accident rate, productivity would also suffer. Most studies have centred on specific occupational hazards or on extreme conditions and have shown correlation between temperature and time lost through sickness and between temperature and the accident rate. All the studies show that workers are much more sensitive to non-ideal conditions of temperature and air movement than to other environmental factors for which the range of acceptability is very much wider. Experiments devised to give precise numerical expression to standards suffer from the fact that the environment, even under test conditions, is made up of a complex interaction of factors and it is quite impossible to expose the subject to only one factor at a time. However, some guidance is given by studies in which the subject is required to carry out a standardised task and his performance is assessed under varying environmental conditions of temperature, noise, lighting, etc.

Highest level

At this level psychological factors play an important part, i.e. the interplay between the subject's immediate intake of information from his surroundings and that information which is stored in his memory from previous experience. A host of factors determine the probability that a particular stimulus will be transmitted through a sensory channel – the nature of the stimulus, existing events taking place in the sensory system, the arousal level of the individual, previous experience of the stimulus, etc. The incoming signal is transformed from a sound wave, light quanta or heat into a series of neural pulses which travel to the central nervous system. There is an interplay between the incoming data and the information stored in the memory after which a response takes place. The response may be an active one, e.g. distraction by noise, dazzle by light, shivering in a cool room, or a passive response in which the *status quo* is preserved.

Studies in this field show the difficulty of establishing universally acceptable standards at this level of sophistication. Not only will people select different stimuli, but they will interpret them differently according to their past experience. One would expect therefore a fairly wide range of preferences according to the age, sex, race, habits, etc., of the person involved. Thus, rigid controls based on narrow and probably short-lived concepts of what is desirable are inappropriate – the aim should be to provide as wide a range of options as possible. This would also suggest giving the individual user a greater voice in the standard of his immediate environment rather than laying down a uniform standard.

The internal environment

Maintenance is part of the process which has been defined as 'facilities management' and which is concerned with providing an acceptable internal environment. Whereas asset management extends to the overall needs of the estate, facilities management is largely limited to the spaces within individual buildings. The objective is to provide a service which ensures that these spaces can be used as effectively and economically as possible. The emphasis is on an understanding of user needs and the complex interaction between the environment and the efficiency with which a building is utilised.

The main areas covered by facilities management are illustrated in Fig. 2.3. The basic parameters of the space shown in the inner circle are fixed to a large extent by the initial design but may be capable of subsequent modification. The outer ring shows the activities necessary to make these spaces habitable. As these activities are so closely related they are more economically carried out under unified control. Also, each will benefit from the higher status accorded to the larger grouping.

The main environmental sub-systems that will be considered are visual, thermal and acoustic.

Visual environment

Lighting
Good seeing results from a combination of good lighting and good sight. As stated by Hopkinson[5] the ability to recognise

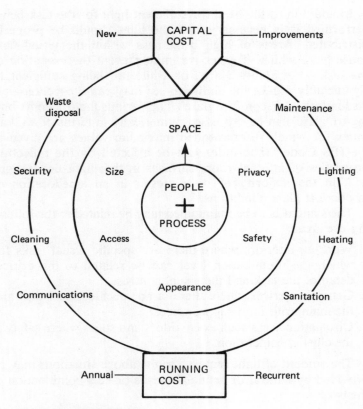

Fig. 2.3 Facilities management

fine detail or visual acuity is dependent not only on the lumin-
ance of the object being viewed and the difference in brightness
between the object and the background but also on the optical
system of the eye, the physical condition of the retina and the
intelligence of the brain in interpreting the visual signals from
the eye. By the use of experimental data it is possible to
prescribe the amount of illumination required for particular
tasks. However, the recommended levels are influenced by
economic considerations and are not absolute values. Thus
illumination levels recommended in different countries vary
between very wide limits, not because of differences in the
quality of eyesight, but because of the cost of lighting and the
standard expected by the users of buildings in those countries.

In addition to the need for sufficient light for the task being performed it is necessary that the light should be properly distributed. Areas of high brightness within the visual field cause glare which, if excessive, may prevent the execution of the task — 'disability glare', or while not being sufficient to significantly reduce the ability to see may cause psychological discomfort — 'discomfort glare'. The various factors contributing to 'glare' can be expressed numerically in terms of a Glare Index for which recommended maximum values are given in the IES Code.[6] The index can be affected by the reflecting properties of internal room surfaces in that light decorations will, in the majority of cases, result in an amelioration of discomfort glare conditions.

The standards to be maintained may be related to the following use areas:

1. Working interiors where there are specific visual tasks for which the illumination level can be related to the critical detail of the task and the visual contrast.
2. General interiors where there is no specific visual task and the main criterion is appearance.
3. Circulation areas such as corridors and stairs where safety is the chief consideration.

The amount of light required in the above situations may be provided by natural or artificial means or be a combination of the two.

Daylighting
Maintenance should be planned to preserve the standard of daylighting provided by the original design. Daylight depreciates for two main reasons:

1. The deposition of dirt on windows reduces the amount of light entering the room.
2. The deposition of dirt on internal room surfaces reduces the internally reflected component of the daylight factor.

The amount of dust and dirt deposited is related to the locality in which the building is situated and the activities which it houses. The rate at which the dirt will collect on a surface will depend on the angle of inclination of the surface, its texture and whether it is exposed to the washing action of rain.

From a knowledge of these factors, it is possible to determine the frequency with which windows must be cleaned to prevent the standard of illumination falling below what is considered to be a desirable level (see Table 2.1).

Table 2.1. Recommended minimum cleaning frequency for windows in various types of building in non-industrial areas. For industrial areas and cities the cleaning interval should be halved (*Source*: Illuminating Engineering Society Report No. 9, 1967)

Type of building	Side windows	Roof lights
Offices	3 months	12 months
Public offices, banks, etc.	2 weeks	3 months
Shops	Outside one week Inside two weeks	6 months
Shops (in main streets)	Outside daily Inside one week	3 months
Hospitals	3 months	6 months
Schools	3–4 months	12 months
Hotels (first class)	2 weeks	3 months
Factories (precision)	4 weeks	3 months
Factories (heavy work)	2 months	6 months
Domestic (by contract)	4–6 weeks	—

If the designer can assume that these frequencies will be adhered to he can allow for depreciation by providing a slightly higher initial standard than that recommended for the particular use area. However, there are situations where the regular cleaning of windows is not possible due to inaccessibility or will not be undertaken because of cost. Cleaning factory roof glazing presents particular problems and is frequently neglected. The result is that the transmission of such glazing in a heavily polluted industrial area may be no more than 10 per cent of its design value. In such cases it would usually be cheaper in the long term to design for permanent artificial lighting since the additional electricity costs would be offset by savings on cleaning and heating arising from the substitution of normal roof coverings for the glazed areas.

Where cleaning is delayed any deficiency in the amount of

natural light would have to be made good by artificial light. The economic periods for window cleaning therefore should be those which minimise the total cost for window cleaning and electric lighting. However, such arithmetical exercises tend to be academic in that there are so many unknown factors in dispute, e.g. it is by no means certain that the artificial lighting would be switched off even if the windows are clean. It is probable that the theoretical savings would only be realised in practice by the installation of automatic controls to adjust the level of artificial lighting according to available daylight.

Artificial lighting
The level of artificial lighting decreases progressively due to the accumulation of dirt on fittings and room surfaces and to the fall in lamp output. The formula normally used to calculate the number of lamps makes allowance for this fall-off by incorporating a maintenance factor:

$$NF = \frac{E_d \times A}{CU \times MF \times r}$$

where N = number of lamps required
    F = lighting design lumens per lamp
    E_d = IES recommended illumination value
    A = area of working plane
    CU = coefficient of utilisation
    MF = maintenance factor
    r = absorption factor of the atmosphere (1·0 for normal situations)

The maintenance factor (MF) is made up of two components – fitting depreciation and room surface depreciation. The basic maintenance factors for selected fittings, building conditions and cleaning intervals for most practical conditions are tabulated in the IES Code.[6]

Lamp replacement
Light output falls off at a comparatively slow rate and therefore replacement policies are usually based on lamp survival.

There are two basic approaches:

(*a*) Spot replacement: the total replacement cost (*C*) per lamp is equal to the cost (L) of the lamp plus the labour cost (S) to replace, i.e. $C = L + S$.

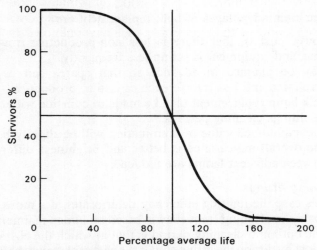

Fig. 2.4. Typical lamp mortality curve (Source: Illuminating Engineering Society Report No. 9, 1967)

(*b*) Bulk replacement: the object is to avoid the costly random use of labour involved in replacing lamps individually. The typical lamp mortality curve (Fig. 2.4) shows that at 80 per cent of the average life the failure rate increases very rapidly and replacement of all the lamps at this time is generally accepted as giving maximum economy.

The cost per lamp is given by the following formula:

$$C = (L + B) + \frac{F}{100}(L + S)$$

where C = total replacement cost per lamp
L = lamp cost
S = labour cost per lamp for spot maintenance
B = total bulk re-lamping labour cost as cost per lamp
F = % of lamps failed at end of bulk re-lamping interval

For example, comparing the costs of the two alternative methods for a tungsten installation assuming that the lamp cost is 10p and the labour costs for spot replacement and bulk replacement 50p and 25p per lamp respectively.

Spot replacement $= 10 + 50 = 60$p per lamp

Bulk replacement $= (10 + 25) + \dfrac{20}{100}(10 + 50) = 47$p per lamp

The main advantages of bulk replacement are:

1. Lower cost in that there is less non-productive travelling time, and equipment is set up less frequently.
2. May be planned in advance so that spares can be bulk purchased and less interference caused to production.
3. Bulk lamp replacement may be made to coincide with one of the fittings cleaning periods.
4. The maintained value of illumination will be slightly better.
5. The overall appearance is better and brightness differences between adjacent lamps are reduced.

Cleaning fittings

In this case the lighting efficiency deteriorates at a more rapid rate and the cleaning frequency is based on lumen maintenance, i.e. the optimum cleaning period is that at which the cost of the light lost by dirt on the fitting equals the cost of cleaning the fitting.

However, where a bulk lamp replacement policy has been adopted the fittings would normally be cleaned at the time of changing the lamps.

Cleaning and decorating room surfaces

It is difficult to determine from purely economic considerations the optimum cleaning and redecoration periods for room surfaces. In some cases, the maximum periods are imposed by statute, but in most cases they would be based on a subjective assessment of what standard is acceptable to the occupants. A degree of objectivity can be introduced by breaking the area down into elements – floors, walls, ceiling, windows, etc. and giving these scores according to their cleanliness, order and repair. Different weightings would be given according to the importance of the element, e.g. in a production area the points may be graded as follows:

Floors	40	
Walls	15	Weights represent amount of time
Ceiling	10	to rectify defective condition
Windows	10	based on standard cleaning times
Lighting fixtures	10	and degree of disfigurement
Equipment	15	

For each element a chart is produced showing the number of points to be deducted for various defects. The total cleanliness rating for the area can then be determined and compared with the standard laid down. The standard is based on what is considered desirable and what is possible under the prevailing conditions and within the budget figure for the area. Thus, a canteen might be assigned a standard of 90 while a warehouse might be given a standard as low as 65 because the condition does not affect production, the health of the people working there, or public relations. The method relies essentially on subjective judgment and its main virtue appears to be that it provides a consistent check list for detailed inspections of the work done.

Thermal environment

The sensation of thermal comfort is determined by the reaction of the individual to the factors governing the loss of metabolic heat. The body is maintained at an essentially constant temperature by automatic reactions that adjust the levels of sensible and latent heat exchanges. A sensible heat exchange occurs by conduction through the skin and clothing and then by convection to the air and by radiation to surrounding surfaces. Latent heat exchange occurs by evaporation of moisture at the skin surface and by the exhalation of warm humid air.

The important factors, therefore, are the temperature of the air and surrounding surfaces, the rate of air movement and the relative humidity. The control of these factors depends largely upon the correct operation and maintenance of the service installations which is outside the scope of this book. However, the condition of the building fabric exerts an important influence on the standards achievable. Damp walls resulting from lack of maintenance will affect the rate of heat loss from the building and the relative humidity while a broken window pane, if not replaced, will certainly affect the rate of air movement. The difficulty lies in quantifying the precise effect of, say, a damp patch of a given size, and one must fall back on subjective assessments of visual tolerance.

Light/Heavy weight bldgs tie in with response and with services

Acoustic environment

Noise or unwanted sound can cause annoyance and distract an individual from his work task, depending upon the loudness and the sensitivity of his hearing system. The source may be:

(*a*) External: traffic, aircraft, factories.
(*b*) Internal: mechanical and engineering services, circulation services (lifts), communication services (telephones), process machines and people.

Building maintenance can assist in many ways to reduce the ambient noise level by repairing or renewing noisy equipment, preventing direct air paths through badly fitting doors or windows, renewing badly worn, resilient floor coverings to reduce impact noise and the avoidance of work which will nullify the effect of features designed to restrict the transmission of sound, e.g. gaps provided to give structural discontinuity.

Value considerations

The value of a building is determined by the demand for the services which it offers in combination with other factors of production. Thus in the absence of demand a building has no value to be maintained and neither the initial cost nor the standard of maintenance has any economic significance. Of course, if the lack of demand is purely temporary and it can be foreseen that there will be a future demand for the building then maintenance expenditure can be related to the anticipated future return discounted to present values.

Assuming, however, that a demand exists, the question arises as to how far the condition of the building would affect the price which would be paid for its services in the open market. This can only be assessed by determining the relationship between building condition and user activities. The difficulty lies in separating the contribution made by maintenance from that of the many other factors, Maintenance interacts with other costs and also with revenue, e.g. for shops it is assumed that a higher level of maintenance will draw more customers and induce them to buy more or pay higher prices, while in manufacturing the standard of maintenance may affect the output of the workers and the quality of the product.

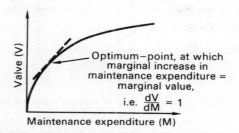

Fig. 2.5. Maintenance expenditure related to building value

It has been proposed that the optimum level of expenditure on maintenance is that which gives the maximum return (see Fig. 2.5). The assumption is that there is an incremental relationship between maintenance expenditure and the value derived from that expenditure and that each additional increment in maintenance expenditure produces a progressively smaller increase in value.

While this would appear a convenient analytical model, it does not accord with reality in that it presupposes a precise and calculable relationship between the standard of maintenance and the efficiency of user activities. It is a matter of common observation that many activities show a high degree of tolerance to the surrounding environment and that quite substantial changes in the condition of the building have little or no effect on them. It is necessary, therefore, to identify:

1. Those user activities which are sensitive to the physical condition of the building and to isolate some aspects of those activities which display measurable changes under different environmental conditions, e.g. speed or accuracy of working.
2. Those building elements which play a significant role in providing the necessary conditions and to assess their probable rate of deterioration and the cost growth of repairs.

Thus, where there is a relationship between building condition and user activities, the consequences of delaying maintenance can be made explicit. In some cases, the loss in value (decrease in user efficiency) may be immediate and progressive, as shown

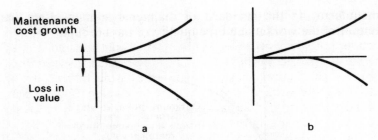

Fig. 2.6. The effect on building value of delayed maintenance

in Fig. 2.6a, while in other cases the loss may not occur until deterioration is far advanced as in Fig. 2.6b. Where the benefits are long-term the expenditure incurred on preventive maintenance is not very different from that incurred in the initial erection and is more in the nature of capital than current expenditure. The major difference is that maintenance does not produce any new capital asset and the measure of its effectiveness lies in the continuing ability of the building to earn a satisfactory rate of return on the total cost, i.e. initial and subsequent maintenance costs.

The aim should be to optimise the total costs as illustrated in Fig. 2.7. The direct costs represent the estimated expenditure on maintenance and the indirect costs the additional costs incurred through lack of maintenance, e.g. extra labour costs, reduced sales, etc. It is assumed that the direct costs are in respect of work which is necessary and which is efficiently

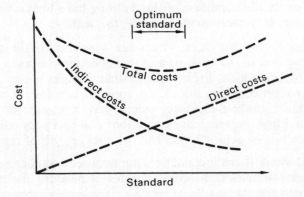

Fig. 2.7. Direct and indirect costs related to maintenance standard

carried out. Ineffective or inefficient work will not alter the optimum standard but it will raise the total cost of achieving that standard. Also it is necessary to take into account the time-scale of events and project the model into the future. Thus where work is deferred the indirect costs should include the future cost of executing the work discounted to present value. This is particularly important in the case of components which are subject to progressive deterioration and which have a steep cost growth.

The model gives only the total cost and it would still be possible to direct resources to the repair of elements which have the least effect on indirect costs. It is necessary, therefore, to analyse each element in this way in order to obtain a series of sub-optimum costs which together would represent the optimum solution for the building as a whole. There is, of course, a danger in rigidly applying financial controls which may result in essential work either not being done or being done inadequately with costly rectification later.

It should be borne in mind that laying down a fixed cost limit will result in a variable amount of work being carried out according to the prices charged by contractors. This will depend not only on the nature of the work and any special difficulties surrounding its execution, but also upon the market conditions prevailing in the locality. Thus, if the price of maintenance were to double, only half the work could be carried out with the possibility of serious long-term consequences.

Also, maintenance costs are determined by quite different factors from production and other activity costs. It is not difficult to conceive of large-scale work which is necessary to preserve the structural stability of the building, but which has no immediate effect on the use of the building, while some comparatively minor item could lead to a production stoppage with very serious financial consequences. Thus, a cost limit related to output would be completely inadequate in the one case and extravagant in the other.

It becomes even more difficult to fix standards in this way where there is a large element of subjectivity involved. For instance, internal painting and decorating forms a large proportion of building maintenance expenditure and yet the cycles are fixed on the basis of supposed visual acceptance. It is unlikely that redecorating an office every five years instead of every four years would have any measurable effect on labour costs, either resulting from a slower pace of working, higher labour turnover or the need to pay higher wages in compensation. One might assume that maintaining a high standard in areas open to the public would create a favourable image of the firm and thus result in indirect benefits. In such cases, the costs incurred would be of a similar nature to those incurred in advertising or public relations and would have to be judged according to similar criteria.

Where the object of the maintenance is to reduce the probability of injury or damage to the occupants or contents the expenditure would be akin to that incurred in insurance and the level would be related to the risks involved. Clearly no building can be perfectly safe in all respects and a proper balance must be achieved between the financial consequences of failure and the cost of work designed to prevent failure.

References

[1] Robertson, J. A. (1969) The planned maintenance of buildings and structures.' The Institution of Civil Engineers, Proceedings Paper 7148S.
[2] Kemp, J. (1966) *Design, Maintenance and Operation.* London: HMSO.
[3] Ray, G. H. (1969) 'Budgeting for maintenance.' Conference on Building Maintenance, London.
[4] Stevens, R. (1973) 'Maintenance standards and costs.' National Building Maintenance Conference, London.
[5] Hopkinson, P. G., Petherbridge, P., Longmore, L. (1966) *Daylighting.* London: Heinemann.
[6] Illuminating Engineering Society (1968) IES Code – Recommendations for Lighting Building Interiors.

CHAPTER THREE

Statutory standards

Legislative controls

The legislative control of building standards is limited for the most part to matters affecting health and safety and may be divided broadly into Acts of Parliament which lay down the basic framework of control and regulations made under such Acts to fill in the detailed requirements. There are a great many Acts of Parliament which relate in one way or another to building standards; some apply to buildings in general while others control the condition of specific types of building. Those which are of most concern to the maintenance manager may be classified as:

1. Those which apply to the design and physical requirements of new construction, additions, alterations and works necessitated by a material change of use. These requirements are largely covered by the Planning Acts and the Building Regulations.
2. Those which are of a continuing nature and apply during the occupation of the building. The most important provisions are contained in the Health and Safety at Work Act and the regulations made thereunder. These will progressively replace those made under the Factories Acts and the Offices, Shops and Railway Premises Act.

Town and Country Planning Acts

The main provisions regarding planning are to be found in

the 1971 Act. The general principle is that a landowner has no right to use his land for any purpose other than its present use unless he first obtains permission for the change of use. The object of the statutory provisions is to control 'development' which is described in the Act as 'the carrying out of building, engineering, mining or other operations in, on, over or under land or the making of any material change in the use of buildings or other land'. However, certain operations are deemed not to involve development and include works of maintenance, improvement or other alterations which affect the interior of the building but do not materially affect the external appearance. Also the Act empowers the Secretary of State to make development orders which exempt the developer from obtaining planning consent for the permitted development. The Town and Country General Development Order, 1981, granted such development consent to the enlargement, improvement or other alteration of a dwelling-house provided there is no increase in its cubic content (measured on the external faces of the walls) exceeding 70 cubic metres or 15% whichever is the greater subject to a maximum of 115 cubic metres and that the height does not exceed the height of the highest part of the roof of the original building or project beyond its front.

Except in cases of permitted development it is necessary for planning permission to be obtained from the local planning authority which may grant it conditionally or unconditionally, or may refuse it.

The Town and Country Planning Acts also provide for the Secretary of State to compile lists of all buildings which are of historical or architectural interest. Buildings are selected on the grounds that they are good examples of a particular architectural style or piece of planning, that they incorporate some technological innovation, that they have associations with well-known characters or events, or for their group values, e.g. squares and terraces. In practice any building dating before 1700 is automatically listed and most of those built between 1700 and 1840. Listed Building Consent must be obtained from the local planning authority for demolition, alteration or extension of any listed building in any manner which would affect its character. Consent is therefore necessary even in the case of such minor alterations as removing the glazing bars from Georgian windows. It is interesting to note that a local authority can now prevent an owner from destroying a listed building through deliberate neglect by serving on him a Repairs Notice specifying the work reasonably necessary to preserve the building. In the event of the necessary work not being carried out the Authority or the Secretary of State may make a Compulsory Purchase Order and pay only minimum compensation. Also where it appears to the local authority that the owner has demolished or altered part of the building in contravention of the Act they may serve on him a Listed Building Enforcement Notice requiring him to restore the building to its former state.

Further controls were introduced by the Town and Country Amenities Act, 1974, which applied listed building controls to all buildings within conservation areas. Thus anyone wishing to demolish a building within a conservation area must apply for listed building consent to the local planning authority either separately or as part of an application for planning permission to redevelop the site. In such cases special regard should be paid to such matters as bulk, height, materials, colour, vertical and horizontal emphasis and grain of design. The Secretary of State has power to make grants for the maintenance and repair of buildings of 'outstanding' architectural or historic interest and for the preservation or enhancement of character or appearance of conservation areas.

Building Act, 1984

This Act consolidates most of the primary legislation relating to buildings, including the relevant parts of the Housing and Building Control Act, 1984, which introduced fundamental changes in the method of building control.

The main provisions in the context of maintenance and improvement work are:

Part I Building Regulations

Section 1 gives the Secretary of State the power to make regulations with respect to the design and construction of buildings and the provision of services, fittings and equipment in or in connection with buildings for the purposes of:

(a) securing the health, safety, welfare and convenience of persons in or about buildings and of others who may be affected by buildings,
(b) furthering the conservation of fuel and power, and
(c) preventing waste, undue consumption, misuse and contamination of water.

Section 2 provides that Building Regulations may impose continuing requirements on the owners and occupiers of buildings including those which were not at the time of their erection subject to regulation. These continuing requirements may apply either to what are called designated functions, e.g. keeping fire escapes clear, or in respect of services, fittings and equipment, e.g. the periodical inspection and maintenance of lifts. In the latter case the requirements may extend to the conditions of use and the submission of condition reports to a prescribed authority. Also under Section 33 it may be necessary to carry out such reasonable tests as may be necessary to ascertain whether or not there has been a breach of a continuing requirement.

Section 6 provides for the issuing of 'approved documents' by the Secretary of State, for the purpose of providing practical guidance on the provisions of the Building Regulations.

Section 7 provides that whilst failure to comply with an approved document does not of itself render a person liable in criminal or civil proceedings, a failure to comply would tend to establish liability whereas proof of compliance would tend to negative liability.

Section 8 gives the Secretary of State power to relax Building Regulations after consultation with the local authority or for this power to be exercised by the authority where the regulations so provide if it is considered that the regulation would be unreasonable in the particular circumstances.

Sections 19 and 20 relate to short-lived materials and the use of materials unsuitable for permanent buildings and provide that a local authority may pass plans showing the use of such materials subject to fixing a period on the expiration of which the building must be removed, or imposing conditions on the use of the building.

Notwithstanding compliance with the Building Regulations a local authority may reject the plans of a proposed building or extension if they fail to show:

> Proper drainage facilities (S21).
> Satisfactory means of access for removal of refuse to a street (S23).
> Adequate fire exits (S24).
> Provision of an adequate supply of wholesome water to a house (S25).
> Sufficient closet accommodation (S26).
> Provision of a bathroom containing a fixed bath or shower and hot and cold water supply (S27).
> Suitable accommodation for food storage to dwellings (S28).
> Freedom of building site from offensive material (S29).

Persons contravening the Building Regulations are liable to be fined and the local authority may issue a 'Section 36 notice' requiring the owner to either pull down the work or, if he so elects, to effect such alterations as may be necessary to secure compliance. An appeal against such a notice may be made under Section 40.

Part II Supervision of building work, etc., otherwise than by local authorities

This part provides for the optional system of supervision of plans and work by an approved inspector and lays down the procedures to be followed in respect of giving the initial notice to the local authority and the final certificate.

Part III Other provisions about buildings

This part gives local authorities the power to require the owners or occupiers of buildings to provide certain facilities or to remedy certain deficiencies. For this purpose the authority may give 'notices' requiring:

1. Provision of a satisfactory drainage system or the execution of such work as may be necessary for renewing, repairing or cleansing drains and appliances (S59).

2. Discontinuance of the use of rainwater pipes for the purpose of conveying soil or drainage from a sanitary convenience (S60).
3. Provision of sufficient closet accommodation (S64).
4. Provision of sufficient and satisfactory sanitary conveniences in a building used as a workplace (S65).
5. Replacement of earth closets where a sufficient water supply and sewer are available (S66).
6. Provision of a wholesome water supply in occupied houses (S69).
7. Provision of sufficient and suitable accommodation for food storage in a house or part of a building occupied as a separate dwelling (S70).
8. Provision of such means of ingress or egress to the building as the authority after consultation with the fire authority deem necessary (S71).
9. Provision of satisfactory means of escape from fire in the case of flats and other designated buildings which exceed two storeys in height and in which the floor of any upper storey is more than twenty feet above ground level (S72).
10. Raising of chimneys where a building is erected or raised to a greater height than an adjoining building (S73).
11. Alteration or closing of cellars and rooms below subsoil water level if constructed without the consent of the local authority (S74).
12. Intention of the local authority to remedy the defective state of a building where the serving of an 'abatement notice' under Sections 93 to 96 of the Public Health Act, 1936, would result in unreasonable delay (S76).
13. Intention of the local authority to take emergency measures where a building is in such a state or is used to carry such loads as to be dangerous (S78).
14. The repair, restoration or demolition of a building which is by its ruinous or dilapidated condition seriously detrimental to the amenities of the neighbourhood (S79).
15. The demolition of a building stating work to be done to safeguard adjoining buildings, e.g. shoring, weatherproofing exposed surfaces, making good any damage caused by the demolition, etc. (S81).
16. Paving and drainage of yards and passages (S84).

In the following cases the authority may apply to a magistrates' court for an order:

1. Temporary order for closing or restricting the use to the public of a building pending the provision of satisfactory means of ingress and egress (S71).
2. Execution of remedial work to dangerous buildings or restrictions on use where danger arises from overloading (S77).

In some cases the person carrying out certain types of work is required to give the authority a prescribed period of notice before starting the work:

1. 24 hours' notice before repairing, reconstructing or altering the course of an underground drain (S61).
2. 48 hours' notice before complying with the requirements of the local authority relating to sealing disconnected drains (S62).

It should be noted that under Section 63 if a water closet, drain or soil pipe is so constructed or *repaired* as to be prejudicial to health or a nuisance, the person who undertook or executed the work is liable on summary conviction to a fine unless he can show that the prejudice to health or nuisance could not have been avoided by the exercise of reasonable care.

Part IV Duties of local authorities

This part sets down the duties of local authorities relating to the serving of notices, powers to enter premises and execute work and procedures for appealing against notices requiring work.

Building Regulations, 1985

The Regulations are made under the Building Act, 1984, and apply in England and Wales to building works and certain changes of use of an existing building.

Building work is defined in Regulation 3 as:

(*a*) the erection or extension of a building

(b) the material alteration of a building
(c) the provision, extension or material alteration of a controlled service or fitting
(d) work required on a material change of use

Building is defined as any permanent or temporary building but not any other kind of structure or erection, and includes part of a building. Certain small buildings and extensions as well as certain buildings used for special purposes are exempt from the Regulations.

The Regulations say nothing about the point at which repair becomes subject to control. Normally repair whether it involves replacement or making good would not be controlled, but where, for example, a whole building has been seriously damaged the repairs could be so extensive that the local authority could reasonably treat the work as the erection of a new building and apply the Regulations.

The Regulations provide for two systems of control, i.e. supervision by the local authority or by a private approved inspector. In the former case the person intending to carry out building work has the option of either depositing full plans (mandatory in the case of offices and shops) or of giving a building notice containing much less detail. The advantage of giving full plans is that if they have been passed by the local authority and the work complies with them the local authority may not subsequently require the work to be taken down or altered. Also where full plans are accompanied by a certificate given by an approved person to the effect that the plans show compliance with certain requirements, e.g. structural stability or energy conservation, the local authority cannot reject the plans on the grounds of non compliance with the requirements to which the certificate relates.

If it is intended that the work should be supervised by an approved inspector the person intending to carry out the work and the inspector should jointly give to the local authority an 'initial notice'. The local authority must accept or reject the initial notice within ten working days but once they have accepted it their powers to enforce the Regulations are suspended. Unless the work consists of the alteration of a one- or two-storey house the approved inspector must be independent of the designer or builder. Unlike the local authority an approved inspector has no direct power to enforce the Regulations but must inform the person responsible for the work that he believes the work is in contravention of the Regulations and if the alleged contraventions are not remedied within three months he is obliged to cancel the initial notice. In that event the local authority becomes responsible for supervision and may request sufficient plans of the work and may require the opening up of work to ascertain if any uncertified work contravenes the Regulations.

The Approved Documents are published separately and give practical guidance on the ways of meeting the Regulations. Other means of satisfying the Regulations may be adopted bearing in mind that in relation to Parts A–K nothing needs to be done beyond that which is necessary to secure reasonable standards of health and safety. Approved Documents may give guidance in the form of technical solutions or acceptable levels of performance. If there is no technical solution which is suitable in the particular circumstances an alternative approach may be adopted based on the relevant recommendations of a British Standard. It should be noted that British Standards and British Board of Agrément Certificates often cover serviceability or recommendations for good practice which go beyond the requirements of the Regulations. Regulation 8 states that 'no obligation imposed by these Regulations to comply with any requirements in Parts A to K shall require anything to be done beyond what is necessary to secure reasonable standards of health and safety for persons in or about the building and others who may be affected by any failure to comply with that requirement'.

The supporting documents which should be read in conjunction with the Building Regulations are:

Manual to the Building Regulations 1985
Mandatory Rules for Means of Escape in Case of Fire

plus the following Approved Documents:

Reg 7 – Materials and workmanship
A1/2/3 – Structure
B2/3/4 – Fire
C1/2/3/4 – Site preparation and resistance to moisture
D1 – Toxic substances

E1/2/3 – Sound
F1/2 – Ventilation
G1/2/3/4 – Hygiene
H1/2/3/4 – Drainage and waste disposal
J1/2/3 – Heat producing appliances
K1/2/3 – Stairways, ramps and guards
L2/3/4/5 – Conservation of fuel and power

Housing Acts

Standards relating to residential premises are laid down in various Housing Acts and relate to:

The condition of the fabric	Repair, stability, dampness, natural lighting and natural ventilation.
The equipment and services	Sanitary fittings, hot and cold water supply, drainage, cooking facilities, artificial lighting, heating installations.
The internal layout	Space for activities and circulation, privacy in houses in multiple occupation.
The quality of the surrounding environment	Air pollution, noise level, open space, traffic conditions.

They are defined in the following ways in the Housing Acts.

Housing Act, 1957, Section 4

(Re-enacted Housing Act 1985 S604)

This Act laid down criteria for determining whether or not a house is unfit for human habitation. A house may be adjudged unfit as a result of the severity of one, or the combined effect of two or more, of the following:

1. *Repair.* The state of repair should not be a threat to the health of, or seriously inconvenience, the occupiers. The internal decorative condition is not taken into account. Unfortunately it is difficult to show a direct causal relationship between disrepair and ill health and the judgment must be purely subjective and based on what is considered to be socially acceptable.

2. *Stability.* There should be no indications of further movement which may constitute a threat to the occupants. Observation over a period of time is necessary to determine whether a crack or other defect is indicative of progressive collapse. However, the danger might be inherent in the design and there might be no prior visual signs that the building is likely to collapse.

3. *Freedom from damp.* Dampness should not be so extensive as to be a threat to health. Dampness is a consequence of lack of repair and the same subjective judgment would have to be made as to whether the extent of the dampness was tolerable.

4. *Natural lighting.* There should be sufficient light for normal activities under good weather conditions without the use of artificial light. No absolute standard is laid down but in the case of Semon & Co. *v.* the Corporation of Bradford 1922 it was decided that the critical level of daylight that separated what was considered to be inadequate from what was considered to be adequate was a sky factor of 0·2 per cent. More recent cases suggest that a higher standard would be required to meet present-day needs.

5. *Ventilation.* There should be adequate ventilation of all habitable rooms and working kitchens to the open air. The Building Regulations require the opening parts of windows to be not less than one-twentieth of the floor area and this may be used as the basis for assessing the adequacy of the ventilation.

6. *Water supply.* There should be an adequate and wholesome supply of water within the house.

7. *Drainage and sanitary conveniences.* There should be a readily accessible WC in a properly lighted and ventilated compartment.

8. *Facilities for preparing and cooking food and for the disposal of waste water.* There should be a sink with an impervious working surface, a piped water supply and cooking appliance.

9. *Internal arrangement.* This was added by the Housing Act, 1969, and requires that the internal layout of the house should not constitute a hazard or cause serious inconvenience, e.g. steps too steep or without landings opening directly off bedrooms, WCs opening directly off living rooms, etc.

Housing Act, 1969, Section 72

This Act introduced the concept of improvement grants and made it possible for a local authority to take action before a house actually becomes unfit. If the local authority is satisfied that substantial repairs are necessary to bring the house up to a reasonable standard, having regard to its age, character and locality, it can serve notice upon the person having control of the house requiring the necessary works to be executed. This does not apply, however, to works of internal decorative repair.

Housing Act, 1974 (as amended 1980)

(Re-enacted Housing Act 1985 S467–498)

This Act provides for local authorities to designate housing action areas (HAAs), general improvement areas (GIAs) and priority neighbourhoods which were discontinued by the 1980 Act. The Act provides for payment of four types of grants:

1. Improvement grants

These used to be called discretionary grants indicating that they are payable at the discretion of the council. Before approving an application the local authority must be satisfied that the premises when improved will attain the 'required' standard. This is defined in Section 61 and requires that the improved dwelling will have all the standard amenities, a useful life of at least 30 years and will:

(a) Be in good repair and substantially free from damp.
(b) Have adequate natural lighting and ventilation to each habitable room.
(c) Have adequate, safe and sufficient artificial lighting and power points.
(d) Have adequate drainage facilities.
(e) Be in a stable structural condition.
(f) Have satisfactory internal arrangements.
(g) Have satisfactory facilities for preparing and cooking food.
(h) Have adequate facilities for heating.

(*i*) Have fuel and refuse storage facilities.
(*j*) Have adequate thermal insulation to the roof spaces.

This list is often referred to as the ten-point standard.

On receipt of the application the authority will inspect the premises and decide whether all the works scheduled qualify for grant aid and if any additional work is needed to meet the conditions. If a grant is given it will be the appropriate percentage of the estimated approved expenditure up to a stated maximum (the eligible expense limits are up-dated from time to time to take account of inflation and the reader is advised to obtain current figures from the local authority). The appropriate percentage is normally 50 per cent but this may be up to 75 per cent in HAAs (cases of special hardship up to 90 per cent) and 65 per cent in GIAs. In assessing the figure of eligible expense not more than half of the total estimate may be in respect of repairs and replacements. Grants are only available to owners and occupiers of premises with rateable values below certain limits.

2. Intermediate grants

These are available for the provision of standard amenities and were previously called standard grants. The relevant amenities include:

(*a*) A fixed bath or shower.
(*b*) A hot and cold water supply at a fixed bath or shower.
(*c*) A wash-hand-basin.
(*d*) A hot and cold water supply at a wash-hand-basin.
(*e*) A sink.
(*f*) A hot and cold water supply at a sink.
(*g*) A water closet.

The council has the right to insist that when the work has been completed the dwelling will be 'fit for human habitation' (see 1957 Act) although in special cases they may waive this requirement. These grants are mandatory and may include the cost of associated repairs up to the approved maximum.

3. Special grants

These are available for the provision of standard amenities

and means of escape from fire including associated repairs in houses in multiple occupation.

4. Repair grants
These are available only for major structural works to houses or flats built before 1919. Such grants are normally discretionary except where a 'repairs notice' has been served when the council must give a repairs grant at a maximum rate.

Grants are also available for loft insulation under the Government's Home Insulation Scheme 1978.

| | Limits of eligible expense (as at May 1983) | |
	Greater London	Elsewhere
	£	£
Improvement grants		
priority cases	13 800	10 200
(houses in particularly bad condition and houses in HAAs)		
non-priority cases	9000	6600
Intermediate grants		
standard amenities (total)	3005	2275
repairs element	4200	3000
Special grants		
standard amenities	depends on number of amenities needed	
means of escape from fire	10 800	8100
repairs element	4200	3000
Repairs grants	6600	4800

Note – The maximum grant is the percentage of the above cost allowable by the council up to the maximum rate which applies to the particular case.

(The aforementioned housing provisions were consolidated in the Housing Act, 1985.)

Defective Premises Act, 1972

The purpose of the Act is to impose duties in connection with the provision of dwellings and to amend the law as to liability for injury or damage caused to persons through defects in the state of premises.

Duty to build dwellings properly

A person taking on work for or in connection with the provision of a dwelling (whether by the erection or conversion or enlargement of a building) owes a duty to see that the work which he takes on is done in a workmanlike or professional manner with proper materials and that the dwelling will be fit for habitation when completed. The duty is owed to the person for whom the dwelling is provided and to every person who subsequently acquires an interest in the dwelling. Thus the duty is imposed upon builders, subcontractors, architects and surveyors who take on work of the type described and upon developers and others who arrange for builders to take on such work. As this is a statutory duty it cannot be excluded by terms in the building contract. Further, the duty is owed to persons who acquire an interest in the dwelling and were not parties to the original contract.

Where, however, a person takes on work of this nature on the condition that he is to do it in accordance with instructions given by or on behalf of another person he will be relieved of his obligation to the extent that he has complied with the instructions and provided that he does not owe a duty to warn the other person of defects in the instructions and has failed to do so. A person shall not be treated as having given instructions merely because he has agreed to the work being done in a specified manner with specified materials to a specific design. For example, where a builder agrees to provide a dwelling in accordance with plans and specification of his own he does not escape the statutory duty by merely adhering to the agreed plans and specification. No action under this provision may be brought by a person having or acquiring an interest in a dwelling where rights in respect of defects are conferred by an approved scheme.

Duty with respect to work done on premises not abated by disposal

The Act provides that where work of construction, repair or demolition or any other work is done on or in relation to the premises, any duty or care owed in consequence of the doing of the work to persons who might reasonably be expected to be affected by defects in the state of the premises created by the doing of the work, is not to be abated by the subsequent disposal of the premises by the person who owed the duty.

A person who does work to any premises is under a duty at common law to take reasonable care for the safety of others who might reasonably be expected to be affected by defects in the state of the premises arising from such work. Hitherto, however, his liability was extinguished on disposal of the premises. A purchaser or tenant taking premises in such circumstances was affected by the doctrine of *caveat emptor* and could not claim for negligence against the vendor or lessor. The Act now removes this special immunity of vendors and lessors from liability for negligence.

These provisions do not apply:

(a) Where the disposal was a letting and the 'tenancy' commenced before 1 January 1974.
(b) In the case of premises disposed of in any other way, if the disposal was completed or a contract for the disposal was entered into before 1 January 1974.
(c) If the relevant transaction disposing of the premises is entered into in pursuance of an enforceable option by which the consideration for the disposal was fixed before 1 January 1974.

Landlord's duty of care

Where premises are let under a 'tenancy' which puts on the landlord an obligation to the tenant for the maintenance or repair of the premises, the landlord owes to *all persons* who might reasonably be expected to be affected by defects in the state of the premises, a duty to take such care as is reasonable in the circumstances to see that they are reasonably safe from 'personal injury' or from damage to their property caused by the 'relevant defect'. The duty is owed if the landlord knows (whether as a result of being notified by the tenant or otherwise) or if he ought in all the circumstances to have known of the relevant defect. The term 'relevant defect' means a defect in the state of the premises existing at the material time, i.e. 1 January 1974 for tenancies entered into before that date or the date when the tenancy commenced, and arising from or continuing because of a failure by the landlord to carry out his repairing obligations.

This provision replaces Section 4 of the Occupiers' Liability Act, 1957, and extends to all those who might reasonably be expected to be affected by the defect (e.g. a passer-by on the highway or a neighbour in his garden), the duty of care previously owed merely to 'visitors' to the premises.

Where the premises are let under a 'tenancy' which expressly or implicitly gives a landlord a right to enter premises to carry out maintenance and repair work then, as from the time when he first is, or by notice can put himself, in a position to exercise the right he is to be treated for the purposes of this provision as if he were under an obligation to the tenant for that description of maintenance or repair of the premises.

Factories Act, 1961, and Offices, Shops and Railway Premises Act, 1963

The detailed legislation made under these statutes will continue in force until superseded by new regulations under the Health and Safety at Work Act.

Factories Act, 1961

The expression 'factory' is defined as any premises in which persons are employed in *manual labour* in any process for, or incidental to, any of the following:

(a) the making of any articles or part of an article;
(b) the altering, repairing, ornamenting, finishing, cleaning or washing or breaking up of any article;
(c) the adapting for sale of any article;
(d) the slaughtering of cattle, horses and certain other animals;
(e) the confinement of such animals at certain premises while awaiting slaughter;

being premises in which the work is carried on in the way of trade or for purposes of gain (S175(1)).

The duty of observing the provisions of the Act falls primarily on the occupier, although the owner will incur some liability in certain cases where parts of the same building are let as separate factories. Also a person employed in a factory has a duty to use the means and appliances provided for his health and safety and must not wilfully misuse them.

Offices, Shops and Railway Premises Act, 1963

Office. A building or part of a building principally used for office purposes. Office purposes include administration, clerical work, handling money and telephone and telegraph operating.

Shop. A building or part of a building in which the sole or principal use is the carrying on of retail trade or business.

Railway premises. A building occupied by railway undertakers for the purposes of the railway undertaking carried on by them and situated in the immediate vicinity of the permanent way.

The duty of observing the provision of the Act falls primarily on the occupier, although the owner is liable for observing certain of the provisions where a building is let to a number of separate occupiers or where only part of the building is leased to an occupier. In such cases the owner is liable for the following:

Common parts. The owner has a duty to ensure that these are kept clean and suitably lighted and that the floors, stairs and passages are of sound construction and properly maintained.

Sanitary conveniences. The owner is responsible for their provision but not for cleaning.

Fire precautions. The owner is responsible for providing fire precautions in all parts of the building but the occupier has a duty to take certain precautions in his own premises.

Administration

The Health and Safety Commission is now responsible for making arrangements for the enforcement of the relevant statutory provisions except where regulations place this duty on local authorities or other bodies. The enforcing authorities are empowered to appoint inspectors who have very wide powers to enter premises and to make such investigations as may be necessary. These may include taking away samples of substances or the atmosphere in premises, questioning any persons who might assist in their investigations and inspecting any books or documents which are required to be kept by a statutory provision or regulation.

Comparison of provisions of Factories Act (FA) and Offices, Shops and Railway Premises Act (OSRPA)

The general provisions of these two statutes are broadly similar and, to avoid repetition, will be treated together in a comparative way. They are concerned with the safety and welfare of employees working in the building and do not extend to the safe keeping of the contents. The main provisions which affect maintenance are tabulated below.

	OSRPA	FA
Cleanliness	Premises must be kept in a clean state and no dirt or refuse allowed to accumulate. Floors and steps must be cleaned at least once a week by washing or other means. **S4**	Every factory to be kept in a clean state and accumulations of dirt removed daily. Floor of every workroom to be cleaned at least once every week by washing or other means. All walls and ceilings must (i) where they have a smooth surface be washed or otherwise cleaned every 14 months; (ii) where they are painted be repainted at least every 7 years and washed every 14 months; (iii) in other cases to be whitewashed or colour-washed at least once every 14 months. **S1**

	OSRPA	FA
Cleanliness (*contd.*)		Note: the periods stated are maxima and clearly in some cases, e.g. food factories, much more frequent treatment would be necessary.
Overcrowding	Overcrowding should not be such as to cause a risk of injury to health. The room should be of such a size as to allow 40 ft^2 of floor space per person or, where the ceiling is lower than 10 ft, a minimum of 400 ft^3 per person.　**S5**	As opposite. Every person must be allowed a space of at least 400 ft^3. In calculating this space, no space more than 14 ft above the floor shall be taken into account.　**S2**
Temperature	Provision must be made for the maintenance of a reasonable temperature. Where severe physical effort is not involved a temperature of not less than 16°C (60·8°F) after the first hour is deemed reasonable.　**S6** 　Note: although only a minimum temperature is stated there is some evidence to suggest that high temperatures cause greater discomfort and have an equally detrimental effect on output. Certainly in some of the newer office buildings with large areas of glass the avoidance of excessively high temperatures can be a much greater problem than making good any deficiencies in this respect.	Generally as opposite but regulations may be made to establish a standard of reasonable temperature for any factory.　**S3**
Ventilation	Effective and suitable provision must be made for securing the ventilation of every room	As opposite but subject to any special regulations relating to the particular type of factory.　**S4**

	OSRPA	FA

Ventilation (*contd.*) **Lighting**	by natural or artificial means. **S7** Provision must be made for suitable and sufficient lighting either by natural or artificial means.
	
	    Windows and skylights used for lighting must, so far as is reasonably practicable, be kept clean and free from obstruction but may be whitewashed to mitigate glare.
	
	    All apparatus for producing artificial light must be properly maintained. **S8**
		As opposite but subject to any special regulations in the particular class of factory. **S5**
		    N.B. The duty to provide suitable and sufficient lighting includes a duty to ensure that the lighting is in fact turned on at the appropriate times.
		    In Thornton *v.* Fisher & Ludlow Ltd. (1968) C.A., the defendant company were held to be liable for injuries sustained by a cleaner who tripped up over a coil of wire on a factory roadway. Although there were sufficient and suitable lights along the roadway they were not switched on.

Sanitary conveniences	Provision of suitable and sufficient sanitary conveniences at convenient accessible places. They must be adequately lighted and ventilated and properly cleaned and maintained. **S9**	As opposite but subject to regulation as to what is sufficient and suitable provision for any class of factory. **S7**
Washing facilities	Provision of suitable and sufficient washing facilities with running hot and cold or warm water. Effective means of lighting and proper cleaning of all apparatus.	As opposite. **S58**
Drinking water	Provision of an adequate supply of wholesome drinking water. **S11**	As opposite. **S57**
Accommodation for clothing	Suitable and sufficient provision for clothing	As opposite. **S59**

	OSRPA	*FA*	
Accommodation for clothing (*contd.*)	not worn during working hours to be hung up or otherwise accommodated. Where reasonably practicable arrangements to be made for drying. **S12**		
Seating arrangements	Suitable facilities to be provided for seating where persons have reasonable opportunities for sitting without detriment to their work. **S13**	As opposite.	**S60**
Eating facilities	Where employees eat meals on premises suitable and sufficient facilities for eating must be provided. **S15**		
Floors, passages and stairs	Must be soundly constructed and properly maintained and so far as is reasonably practicable kept free from obstructions and slippery substances.	As opposite.	**S28**
	A substantial handrail must be provided – if the staircase has an open side the handrail must be on that side or where there is a particular risk of accidents it must be on both sides. **S16**	As opposite. **S28** Also there must be safe means of access so far as reasonably practicable to every workplace which should be made and kept safe for any person working there. **S29**	
First aid	This now dealt with by the Health and Safety (First Aid) Regulations 1981 SI 917 made under the Health and Safety at Work Act. The regulations are of general applicability and impose a duty on	As opposite. **S61**	

	OSRPA	FA
First aid *(contd.)*	employers to make proper provision for first aid equipment and for the training of employees in first-aid treatment. Also the Notification of Accidents and Dangerous Occurrences Regulations 1980 SI 804 requires employers to report major injuries and keep records of accidents.	

The terms 'properly maintained' and 'reasonably practicable' occur frequently in the Acts and require explanation:

Properly maintained. For the purpose of the Factories Act the words mean 'maintained in an efficient state, in efficient working order, and in good repair'. The adjective 'properly' would suggest a normal accepted standard of good workmanship with particular emphasis on safety.

Reasonably practicable. The accepted legal definition is that given by Lord Justice Asquith in the case of Edwards *v.* National Coal Board 1949: 'Reasonably practicable is a narrower term than physically possible and seems to me to imply that a computation must be made in which the nature of the risk is placed on one scale and the sacrifice involved in the measures necessary for averting the risk (whether money, time or trouble) is placed on the other. If it be shown that there is a gross disproportion between them – the risk being insignificant in relation to the sacrifice – the defendants discharge the onus placed upon them.'

Fire Precautions Act, 1971

The object of the Act is to ensure the provision of adequate means of escape and related fire precautions in places of public entertainment and resort and in certain types of residential premises.

Premises affected

Premises are brought within the scope of the Act either following the issue of a designating order or when the Secretary of State makes regulations applying fire certificate provisions to the premises or when the fire authority issues a notice relating to a particular dwelling.

The designating order must fall within the following classes of use in order to define a type of building for which a fire certificate is required:

(a) Sleeping accommodation.
(b) Institution providing treatment or care.
(c) Entertainment, recreation or instruction, or for the purpose of a club, society or association.
(d) Teaching, training or research.
(e) Any other purpose involving access to the premises by the public whether on payment or not.

The first designation order was in respect of hotels and boarding houses and came into effect on 1 June 1972. This brought under control premises used for the business of a hotel or boarding house where:

(a) sleeping accommodation is provided for more than six persons being staff or guests, or
(b) some sleeping accommodation is provided for staff or guests above the first floor, or
(c) some sleeping accommodation is provided for staff or guests below the ground floor.

A more recent order which came into effect in January 1977 designated factories, offices, shops and railway premises for which fire certificates are necessary except where not more than 20 persons are employed or not more than 10 persons elsewhere than on the ground floor. Fire certificates are also required for certain premises in which explosive, inflammable or other hazardous materials are stored or in which hazardous processes are carried out.

Observance of the Act

The primary responsibility lies with the occupier of the appropriate part of the building although the fire authority can impose requirements on others. Application must be made to the fire authority for a fire certificate within the stipulated period following publication of the designating order. In the case of dwellings the person concerned is notified direct.

Following inspection of the premises the fire authority will issue a certificate specifying:

(a) The use or uses of the premises which it covers.

(b) The means of escape in case of fire.

(c) The method of ensuring that means of escape can be safely and effectively used at all material times.

(d) The means of fighting fire for use by persons in the building.

(e) The means of giving warning in case of fire.

Additionally, the fire certificate may impose requirements relating to:

(f) Maintenance of the means of escape and keeping them free from obstruction.

(g) Maintenance of other fire precautions specified in the certificate.

(h) Training of staff and keeping records.

(i) Limitation on the number of people who may be in the building at any one time.

(j) Any other relevant fire precautions.

Appeals against decisions of the fire authority on such matters as refusal to issue a fire certificate or work to be done before a certificate will be issued must be made to a magistrate's court within 21 days of the decision being known.

Administration

The fire authorities are the main enforcing bodies, the principal exception being Crown premises which are dealt with by HM Inspector of Fire Services. Inspectors may enter premises at reasonable times and make such enquiries as they think necessary in pursuance of their duty. They will also be responsible for reviewing fire certificates from time to time and making such amendments as are necessary in consequence of material alterations or changes of use of the building.

Health and Safety at Work Act, 1974

This Act provides a comprehensive and integrated system of law governing the health, safety and welfare of workpeople and the health and safety of members of the public who are affected by work activities. Regulations made in pursuance of the Act will progressively replace the current health and safety legislation described earlier in this chapter. However, it is anticipated that the process will take a number of years to complete.

The Act is in three parts: Parts I and II, relating to safety and health, are administered by the Department of Employment; Part III, which extended the scope of the Building Regulations, has been incorporated in the Building Act, 1984.

Part I

The provisions of this part impose general duties in relation to health and safety on the following persons:

1. An employer is to ensure, so far as is reasonably practicable, the health, safety and welfare at work of his employees.
2. Both employers and self-employed persons are to conduct their undertakings in such a way as to ensure, so far as is reasonably practicable, that persons not in their employment are not exposed to risks to their health or safety.
3. A person in control of non-domestic premises which are used by persons not in his employment either as a place of work or as a place where they may use any plant or substance provided for their use must ensure, so far as is reasonably practicable, that the premises and any plant or substance therein is, when properly used, safe and without risk to health. In addition there is a duty to use the best practicable means to prevent the emission into the atmosphere of noxious or offensive substances.
4. A person who designs, manufactures, imports, installs or supplies any article for use at work must ensure that the article is safe and without risk to health when properly used, and is required to carry out such tests as are necessary and to take such steps as are reasonably practicable to ensure that adequate information is available concerning the correct use of the article.

5. An employee must take reasonable care for the health and safety of himself and of other persons who may be affected by his actions and shall not interfere intentionally with or misuse anything provided in the interests of health, safety or welfare.

Regulations may be made describing the circumstances in which safety representatives may be appointed or elected from among the employees to represent them in consultations about health and safety. In certain circumstances the employer may have a duty to establish a safety committee to keep the health and safety measures under constant review.

This part of the Act also provides for the creation of the Health and Safety Commission and the Health and Safety Executive. The Commission has general responsibility for the work of its Executive which is the main body responsible for enforcing the statutory requirements on health and safety. Enforcement powers are also given to local authorities and certain other bodies.

The Health and Safety Commission consists of a chairman with a minimum of six and a maximum of nine other members appointed after consultation with employer and employee organisations, local authorities and other organisations. It is responsible for the continuous task of preparing proposals for the revision and extension of statutory provisions on health and safety and for making appropriate arrangements for research and training.

The Health and Safety Executive consists of three persons appointed by the Commission with staff drawn from the previous health and safety inspectorates for factories, mines and quarries, alkali works, etc. Initially the Executive will continue to administer the provisions of the existing legislation but with additional powers to issue improvement and prohibition notices. Thus, if an inspector is of the opinion that a person is contravening a statutory provision he may issue an 'improvement notice' requiring the person to remedy the contravention. If he considers that the activity involves a risk of serious personal injury he may issue a 'prohibition notice' requiring the cessation of the activity.

Part II

This part re-enacts with amendments provisions previously contained in the Employment Medical Advisory Service Act, 1972. The functions of the Employment Medical Advisory Service are to advise on matters concerning the safeguarding and improvement of the health of employed persons or of persons seeking or training for employment and to give such persons information and advice. The responsibilities are discharged by doctors appointed by the authority to act as employment medical advisers.

Observance of the Act

The maintenance manager has much of the responsibility for protecting the interests of the occupier in providing safe working conditions, safety of access and health and welfare facilities. It should be noted that the Act places a responsibility on the person in control of the premises to ensure that not only are the premises themselves safe but also any plant or equipment within the building. This imposes a duty on the person in charge to ensure that the plant and work methods used by outside contractors are not prejudicial to the health and safety of employees or to others who have rightful access to the building. Thus the maintenance manager could incur a personal liability for injuries caused by such contractors. Also, there is likely to be an increase in the amount of work which has to be done in response to requests from safety representatives.

The safety duties imposed by the Act may be summarised as follows:

1. To make and keep all work places safe S2(2)(d) and (e).
2. To ensure the safety of machines and materials S2(2)(a) and (b).
3. To plan and use safe working systems S2(2)(a) and (e).
4. To train, inform and direct employees S2(2)(c) and (3).
5. To receive and consider employee views S2(6) and (7).
6. To ensure the safety of subcontractors and visitors S4(1) and (2).
7. To protect the wellbeing of co-tenants, neighbours and the public at large S3(1).

8. To test and supply in safe condition any goods for use by employees or others S6(1), (2) and (3).
9. To effectively safeguard all employees wherever they may work S2(1).
10. To set out in writing the arrangements and organisation to achieve the above.

Enforcing measures include:

1. Criminal prosecution of the employer, controllers of premises, designers, manufacturers and suppliers of commodities used at work.
2. Prohibition or improvement notices.
3. Industrial tribunal complaint in respect of 5.
4. Civil claims based on alleged negligence or breach of statutory duty.

It should be noted that where a statutory provision requires certain things to be done it is no excuse for the employer to claim that it was not practicable unless the provision states otherwise. Also it is not possible to escape the liability by delegation to others. The only exception is where the employer has delegated the duty to carry out a statutory provision to someone for whose protection the provision was imposed and he is injured through his own negligence, i.e. the employer may claim contributory negligence.

Control of Pollution Act, 1974

This Act deals with waste on land, pollution of water, noise and pollution of the atmosphere. The part which is of greatest relevance to the maintenance manager is that dealing with noise. Where a local authority considers that noise amounting to a nuisance exists, or is likely to occur or recur it can serve on the person responsible a notice requiring the noise to be abated and if necessary the execution of work specified in the notice. In the case of construction sites the local authority may serve a notice imposing requirements as to how the work should be carried out and the maximum permitted noise level. Regulations may also be made for the purpose of reducing noise from plant and machinery and the emission of atmospheric pollutants from furnaces and engines.

CHAPTER FOUR

Planning

Objectives of planning

The object of planning is to ensure that work considered necessary is carried out with maximum economy, i.e. that the work done satisfies the criteria for effectiveness and efficiency. Robertson [1] defines work as 'cost effective' if it is work required as a result of fair wear and tear and is done adequately ('adequately' presumably refers to both quality and cost). Work not required or work done inadequately is regarded as 'ineffective' in that it produces no benefits and may attract penalties in the form of additional future costs. Another source describes effectiveness as a measure of actual performance against planned performance. This is too limiting in that effectiveness in the context of the promotion of the overall aims of the organisation is dependent upon the adequacy of the plan. If the plan is in itself defective, then it is difficult to see how fulfilling it can be regarded as effective.

A particular problem that has confronted local authorities and other large organisations is that lack of co-operation between client departments has resulted in large building projects from different sectors of the authority coming to readiness at the same time causing bunching of tenders and lack of continuity of work for the DLO. Ideally local authorities should be able to plan over a long period to provide the steady flow of work necessary for the construction industry. Unfortunately uncertainty over the level of government grants and abrupt changes in policy tend to militate against this approach. However there are advantages in involving the DLO in inter-departmental planning so as to provide an integrated rolling programme of building, improvements and maintenance. This approach has been described as 'corporate planning'. Another cause of bunching is the well-known one of attempting to spend the budget allowance before the end of the financial year in case approval is not given for the next financial year. This calls for a more flexible approach to budgeting.

Sources of ineffective costs

1. Unnecessary work including:

   (a) Work over and above that required to maintain the building to the specified standards.
   (b) Making good the effects of neglect, improper maintenance and misuse of the building.
   (c) Rectification of design defects and faulty workmanship in the initial construction.

   Harper[2] suggests that rectification is the most profitable point at which to reduce maintenance costs in that it is avoidable.

2. Uneconomic work resulting from:

   (a) Non-productive time caused by excessive travelling from job to job, waiting for instructions and materials, failure to gain access to premises, inclement weather, etc.
   (b) Improper work methods resulting in more time being spent on the job than necessary and/or waste of materials. The cause may be attributed to imprecise instructions or incompetence on the part of the operative.
   (c) Lack of motivation on the part of the operatives.
   (d) Inappropriate tendering procedures and contract arrangements in relation to the type of work and prevailing market conditions.
   (e) Changes to the nature and scope of the work after commencement.
   (f) Lack of an efficient system of recording and controlling costs.

3. Inadequate work resulting from:

(*a*) Failure to identify the true cause of the defect and to specify the correct remedial work.

(*b*) Improper execution of work due to lack of proper supervision, instructions or operative skills.

(*c*) Lack of adequate safeguards in the contract to ensure that the work is carried out in accordance with instructions and the provision of suitable remedies in the event of non-compliance.

Components of planning system

The essential feature of a planned maintenance system is that failures are anticipated and appropriate procedures devised for their prevention or rectification. It involves having a planned course of action for dealing with the inevitable consequences of deterioration. The plan should be all-embracing and lay down measures for dealing with even remote possibilities, e.g. damage caused by severe weather conditions. Whether or not spares are held in store for the rectification of such defects will depend upon the degree of risk involved and the likely effects of delay on user activities.

The question that is sometimes asked as to what are the advantages of a planned approach can best be answered by considering the consequences of an *unplanned* approach. That is, to do nothing until a defect is reported and then request a contractor to do something about it, often with very little control over the work that he does or the price that he charges. This may be described as the reactive or day-to-day approach in that each day presents a fresh set of completely unforeseen problems that have to be dealt with on a purely *ad hoc* basis. It is in fact the antithesis of planning. The following comparison (see p. 109) highlights the relative merits and demerits of the two approaches.

Of course, not all jobs can be predicted with sufficient certainty for inclusion in a long-term programme. The object should be to obtain the most economic balance between day-to-day and programmed work. This can be determined by a statistical analysis of the frequency with which particular defects occur and whether or not it would be cheaper to renew all suspect components at the same time rather than carrying

	Day-to-day	Programmed
Characteristics	Isolated locations Small scale Low value Unpredictable	Jobs grouped Large scale High value Predictable

Consequences	High % lost time
	Access difficulties
	High administrative costs
	Materials in small
	  quantities
	Difficult to control
	  time, cost and quality
		Minimum lost time
		No access problems
		Low administrative costs
		Materials in economic
		  batches
		Easier to supervise and
		  control

out *ad hoc* repairs on a piecemeal basis. Also, planning is not confined to long-term programmes but is equally applicable to the short-term organisation of day-to-day maintenance. The key factor in the management of maintenance is obtaining positive control over the work to be undertaken during any period.

Work is input from two primary sources:

1. Work initiated by the maintenance department and consisting of larger jobs planned some time in advance. These jobs, while necessary for the long-term preservation of the building, may not have a high degree of urgency.
2. Work requested by the occupier and consisting for the most part of small jobs which, at least by the occupier, are regarded as urgent.

However, if the volume of small jobs requested by the occupiers is too great the frequent withdrawal of men from programmed work to deal with them will seriously affect the overall efficiency of the organisation.

This particular problem was faced by a local authority where originally all requests for housing repairs came from tenants. The result was that the incoming workload was irregular in quantity, type and location and as a result was difficult to plan and schedule. The effect was that non-productive time was excessively high and often aggravated by the fact that access could not be gained to the premises as lack of planning made it impossible to give advance notice to the tenants. Another danger of relying solely on this method of determining the workload is that tenants will not only vary greatly in their views as to what constitutes a desirable standard, but will tend to report superficial defects rather than more serious but less obvious structural defects.

The authority decided to introduce a system of planned inspections at three-yearly intervals to anticipate tenants' complaints and provide a more uniform and objective standard of maintenance. The success of such a scheme depends upon close liaison with the tenants when carrying out the inspections, so that all except the completely unpredictable repairs can be included and dealt with while the mobile workshop is in the area. However, there are components for which the deterioration rate is such that defects will develop between inspections. If these interfere with the user of the building or constitute a hazard, they will have to be dealt with outside the main programme. Where there is a sufficient number of such jobs, it would be advantageous to have a separate mixed gang solely responsible for emergency work to avoid disrupting the programmed work.

Another local authority uses the repair order issued by the user department as the primary document to initiate general repairs. Work input control is achieved by sorting according to priority and previewing by the foreman, who instructs operatives to carry out the work when the materials are available. The system relies heavily on short-term planning, continuity being ensured by the backlog of work existing at any time. Thus, although the flow of work to the labour force remains uniform, the backlog will vary according to the number of requests made by user departments. Inevitably, this will result in some delay in dealing with user requests and it is necessary to ascertain what is an acceptable period. Clearly there will be a direct relationship between the size of the labour force and the delay period. For housing, it is probable that the acceptable delay period will be less than the economic period and that additional cost will be incurred in obtaining user satisfaction. In the case of commercial and industrial premises it may be possible to relate the delay period to the estimated financial losses resulting from the continuance of the state of disrepair.

Schedule/contingency systems

The maintenance plan must therefore strike an economic and socially acceptable balance between the operation of two complementary and interacting systems – schedule or programmed system and contingency system. These systems and the associated procedures are shown in flow diagram form in Fig. 4.1.

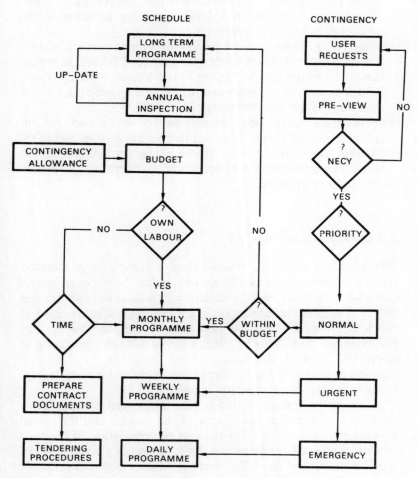

Fig. 4.1. The schedule and the contingency systems of planned maintenance

1. Schedule system

This covers items which tend to deteriorate at a more or less uniform rate and which do not have a high degree of urgency. The procedures take the following forms:

(a) Scheduling work to be carried out at predetermined times. This includes planned preventive maintenance and applies where the incidence of failure can be predicted with some accuracy or where the periods are fixed by statute or contract, e.g. the terms in a lease requiring painting to be undertaken at fixed intervals.

(b) Scheduling inspections to be carried out at predetermined times to detect failures or the imminence of failure. In many cases the exact time of failure is not known with certainty and inspections are necessary to determine whether or not the work is actually necessary.

(c) Scheduling work and inspections to be carried out at predetermined times. This applies where it is possible to predict that certain work will be necessary at a particular time, but an inspection is necessary to determine the extent of any further work.

2. Contingency system

This entails a policy of waiting until a complaint is received from the user before taking action. It has been described as casual maintenance, but the adjective 'casual' is not appropriate in that it suggests an absence of planning. If the frequencies and types of complaint are analysed, procedures for dealing with the remedial work can be planned even though the timing is uncertain. That is, the direct labour force can be sized and manned to deal with the predicted workload and materials kept in store for the more common repairs.

A necessary feature of this system is the need to introduce a delay period between the receipt of the request and the actual execution of the work. This permits the regulation of the flow of work to the labour force and the grouping of similar items of work in space and time. The main difference between schedule and contingency maintenance is that the lead time, i.e. time between notification and execution, is greater for schedule maintenance.

Clearly the longer the lead time the more detailed the preparatory work and the less likely are delays due to lack of precise instructions or non-availability of materials and plant. The most economic policy is therefore one which maximises the amount of schedule work. However, maintenance is a service and there would be little point in producing a least-cost solution which was unacceptable to the user.

Choice of system

The main factors to consider when deciding whether to treat work under the schedule system or the contingency system are:

1. The predictability of failure. Components which deteriorate at a known and fairly uniform rate can be scheduled either for inspection or repair before they actually fail. On the other hand, components which are susceptible to sudden failure, e.g. burst pipes, can only be dealt with under a contingency system. However, incipient faults which may lead to sudden failure may be capable of identification when making the general inspection of the property and work scheduled which will reduce the probability of failure, e.g. lagging exposed pipes.

2. The reporting delay time. This is the time which is likely to elapse between the time when the defect would just be noticed by a qualified inspector and the time when the occupier would report the defect to the maintenance department. It will depend mainly on the inconvenience which the defect causes to the occupier and is not a measure of the seriousness of the defect. If the reporting delay time is less than the economic period for carrying out inspections the work must of necessity be dealt with on a contingency basis. However, it should be noted that in most cases user requests require previewing by someone who is technically qualified to ascertain the cause of the defect and the scope of the remedial work necessary.

3. The rate of deterioration of the component and the corresponding increase in the cost of rectification. This must be considered in conjunction with the reporting delay time and

whether or not it is the type of failure which is likely to prompt an early response from the user.
4. The extent to which the user can be relied upon to report significant defects. This will depend upon the nature of the occupancy and the attitude of the user to the condition of the building.

Factors influencing delay time

As stated earlier, there must be some delay between receipt of a user request and the execution of the work. The factors which determine the permissible delay time include:

1. Safety considerations and, in particular, compliance with statutory requirements. Clearly failures which constitute a hazard to the occupants or to persons coming on to the premises must be given first priority. The penalties for not doing so may be heavy damages in a civil action or prosecution for non-compliance with statutory provisions. Of course, if the penalties are slight in comparison with the cost of doing the work, then non-compliance could be a calculated risk.
2. User satisfaction, which must be considered even though the defect may be in no way dangerous. It is not uncommon to find a difference of opinion between a tenant and a landlord as to what is an acceptable standard. In some cases, maintenance can contribute only a small part in that the basic cause of the dissatisfaction is attributable to inadequate space or lack of amenities. However, the contingency system does have some psychological advantage in that tenants have a greater sense of participation than would be the case if the whole of the work were planned without their intervention. It can, of course, result in some properties being maintained to a higher standard than others according to the initiative taken by tenants in requesting repairs.
3. Effect of failure on the primary activities of the organisation. This would apply particularly to commercial and industrial buildings which, in this context, are similar to plant and equipment, being distinguished mainly by their very much longer life and the lower probability of complete breakdown.

The buildings are factors of production and any savings resulting from delaying the execution of the work must be balanced against the losses suffered by prolonging interference with the carrying out of basic activities. It is necessary, therefore, to determine whether the defect impinges upon the activities of the organisation, e.g. manufacture, sales, care of patients in hospital, etc., and if so to assess the rate of loss incurred from the state of disrepair.

4. Dispersion of job situations. Travelling time is an important component of ineffective costs and may account for 40 per cent or more of the total cost. Where small jobs are widely dispersed, travelling time can be considerably reduced by grouping jobs requiring the same craft skills according to location. The amount of time saved will depend upon the distance from the depot to the job area and the number of jobs that can be executed in one day (see Fig. 4.2).

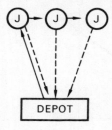 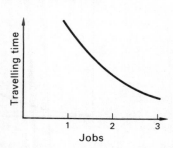

Fig. 4.2. The relationship between job location and amount of work covered

For relatively concentrated areas remote from the main depot, such as scattered housing estates, a similar reduction could be obtained by setting up sub-depots in convenient positions or by the use of mobile workshops. In both cases the additional cost would have to be justified in terms of the saving on travelling costs and quicker service.

Even for single buildings, if of a sufficient size, some grouping of jobs may be advantageous in reducing the costs of conveying plant and equipment to the job area and of setting up for the jobs and clearing away on completion.

5. Cost growth. In most cases the longer a defect is left unattended, the more expensive the remedial work, not only to the component initially affected but also to the surrounding

parts of the structure. It is not possible to lay down any universal rules and in each case the deterioration characteristics of the affected component must be considered as well as the properties of adjoining components. The cause and effect may be widely separated, e.g. water carried along hidden and devious channels within the structure from one part of the building to another. Deterioration profiles for individual elements would be of only limited value in this respect. Ideally they should indicate the effects of failures on adjoining elements and this will depend upon the unique combination of elements for the building under consideration. Figure 4.3 indicates the general pattern.

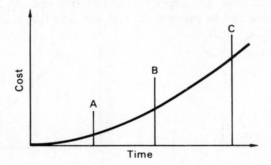

Fig. 4.3. Cost growth related to repair delay
A time when defect just apparent
B time when defect likely to be reported
C time of repair

6. Misuse of property. There is little doubt that buildings in a dilapidated state are not used with the same care as those which are maintained to good standards. It is recommended in BRE Digest 132 (1971) that the basic approach should be to maintain to such a standard as to produce a psychological resistance to vandalistic tendencies. Certainly the maintenance manager should be aware of the temptation provided by a lack of repair and the tendency for faults to escalate as a result of misuse. Also, good feedback to the designer is essential if mistakes are not to be perpetuated.

Programming problems

The characteristics of maintenance render the accurate and comprehensive long-term programming of operations impracticable. This is not to say that such programmes do not serve a useful purpose, but that the uncertainties inherent in a large proportion of the work should be recognised and sufficient flexibility built into the programme to permit inevitable modifications. It was remarked in a report by the Tavistock Institute on the construction industry [3] that such programmes can only be based on assumptions about the variety, quantity and timing of the future application of resources and that persistence in unreal assumptions has led to the often uncritical and inappropriate application of techniques of 'scientific management'. In point of fact, the report was referring to new construction in which the uncertainties are generally very much less than in maintenance. As noted by Skinner[10] such planning systems must be related to the reality of the work actually being done on the buildings and this requires detailed feedback on previous maintenance jobs in order to establish their distribution and frequency. The main factors which militate against accuracy when programming maintenance work are:

1. The small scale and diversity of a large proportion of the total workload. A study[4] of housing maintenance found that there were about 50 requests per week per 1000 dwellings for small repairs costing on average less than £30. The precise identification of such small jobs in a long-term programme is not possible nor would it be economically worthwhile if it were possible. However, the aggregate amount of these jobs over a period of time shows some constancy and a broad classification according to trade, size and frequency is possible.
2. The need to determine the best work sequence for a large number of interlocking tasks involving several trades. Thus, where a number of jobs are proceeding simultaneously on different sites the individual job programmes must be timed in relation to one another so that operatives with the requisite skills are available at the right time on each site. The position is aggravated if the work in a particular trade is discontinuous, necessitating two or more visits to the site.

Jobs which are interdependent in that they rely on the same operative carrying out a critical operation at a specific time are particularly susceptible to delays. If the operative is prevented from performing his task on one job, he may not be available when required on other jobs and the initial delay which may have been quite small escalates throughout the system. One solution is the training of operatives in a number of related trades, so that time spent in waiting on other trades is reduced to a minimum.

3. The uncertain work content. In many cases, particularly when dealing with old buildings, work is uncovered which could not have been foreseen. This does not necessarily imply incompetence on the part of the inspector, who is usually required to reach a decision on the basis of a visual inspection of the surface effects of the fault. Often it is only when the outer coverings have been removed that the full extent of the damage is revealed. In such cases, changes arising out of a greater knowledge of the defect tend to start a chain of consequential changes, which result in much more extensive work being done than was originally envisaged. Also, this may prompt users to ask for additional work and the initial programme quickly becomes outdated.

4. The dispersal of sites. This should be carefully considered when programming, from the point of view both of supervision and of the need to redeploy men to other jobs in the event of unavoidable holdups. Clearly, the geographical grouping of jobs is desirable and also the provision in the same locality of both indoor and outdoor work for operatives of the same trade groups.

5. Interruptions to the normal progress of the work caused by:

   (a) Withdrawal of men to deal with emergencies or to carry out a critical task on some other job.
   (b) Inclement weather.
   (c) Non-availability of essential materials or plant.

6. Irregularity of user requests and the extremely short lead time for executing emergency repairs.

In view of the uncertainties, programmes must be formulated at different levels and constantly revised according to the new information which is fed into the system. The levels which can be identified are:

Long-term   Quinquennial or longer
Medium-term   Annual
Short-term   Monthly, weekly, daily

Long-term programmes

The object of the long-term programming is not so much to lay down the precise dates when work is to be carried out, as to provide a policy framework. The Woodbine Parish Report on hospital buildings recommends a broad survey of each individual building to establish the major non-recurring items which are likely to require substantial sums of money within the next five- to seven-year period. Speight[5] describes this as a broad general appraisal in order to formulate policy.

The purpose of the long-term programme

1. To determine the general level of expenditure on maintenance to achieve the desired standards. Inspections will often reveal that there is a backlog of work to bring the buildings up to the specified standard, and this should be recorded separately.
2. To avoid large fluctuations in annual expenditure by spreading large items and any backlog over a period.
3. To determine the optimum time for carrying out major repairs and improvements, so as not to interfere with the user of the building.
4. To determine the structure and staffing of the maintenance organisation and whether it would be advantageous to employ operatives directly to carry out part or the whole of the work.
5. To gear the maintenance programme to company policy so that it is compatible with decisions relating to the use of the building, e.g. decisions to demolish and rebuild or to move to other premises.
6. To consider the effect of proposed capital works on the maintenance organisation.

Major items of work

The long-term programme will therefore seek to identify the major items of work over the next five to ten years. The information is obtained from past records showing when major repairs were last undertaken and from inspections of the current physical condition of the various elements.

Painting

Usually the programme pivots around the painting and decorating cycles on account of the predictability of this work and its magnitude – about one-third of the total expenditure on maintenance.

In some cases, the lease lays down specific periods for repainting and these must be complied with, even though they rarely represent the most economic cycles.

Where there is sufficient past knowledge of the behaviour of the building and close control is exercised over the quality of paints used and workmanship, it is possible to calculate theoretically optimum cycles. However these periods are essentially averages based on experience and the need for repainting should be confirmed by inspection.

Where appearance is important, it will be necessary to form a subjective assessment as to whether or not the state of the paintwork at the end of the economic period will be acceptable. The difference in cost between repainting at the calculated economic periods and at those judged necessary to maintain a satisfactory appearance will represent the value which the occupier attaches to appearance.

In the case of a large building, each elevation should be considered separately in that being exposed to different weather conditions, the paint films will deteriorate at different rates. Against this must be set any savings that might accrue from having the whole of the external painting carried out at the same time rather than piecemeal.

Internal paint films rarely have any protective function and are subject to very little physical degradation over the periods normally considered acceptable. Any attempt to optimise the cycles must therefore be related to the visual impact on the users and the possible effect on their health and working efficiency. Thus, the frequencies adopted for redecorating the internal spaces vary widely according to the type of user and the importance attached to appearance and cleanliness. This is reflected in the following programme, suggested by Luke[6] for the internal redecoration of hospitals.

Operating theatres	3 months
Wards and kitchens	2 years
Nurses' home	3 years
Offices	4 years
Machine areas	7 years

The work to the various use areas can be phased over a period equivalent to the longest cycle, i.e. seven years in the above example, so that a more or less uniform amount is undertaken each year. Also, it is advantageous to programme at least part of the internal painting to be executed at the same time as the external painting, to provide alternative work in the event of bad weather.

However, the reasons for adopting different cycles should be carefully examined and justified. In a reported case concerning a local authority the adoption of different cycles for painting and redecorating similar buildings was due to the fact that the standards were laid down independently by the sub-committees responsible for administering the various services. It was suggested that a more rational policy could be achieved if the care of buildings was centralised and placed under the control of a single department. A possible argument in favour of retaining the existing system might be that a department concerned solely with the technical aspects of maintaining buildings might overlook the broader issues. Thus, in the case of this particular local authority, additional money was spent on police houses in order to attract officers and men from other constabularies to combat the rising rate of crime in the area. The level of expenditure on maintenance was, thus, just as much a function of the crime rate and availability of police officers in the area as of the physical condition of the buildings. On the other hand, a central organisation would be in a better position to produce a comprehensive rational policy in line with the needs of the individual users.

Major repairs

These usually involve the replacement or renewal of elements or components for the purpose of eliminating areas of high maintenance costs or restoring lost or diminished amenities. Generally the elements involved are characterised by a relatively slow rate of deterioration and an increasing need for minor repairs. Roof tiling falls into this category in that, although defective areas can be repaired, there comes a time when the original tiles have deteriorated to such an extent that it is more economic and functionally more satisfactory to renew the whole of the tiling rather than to continue to patch an increasing number of small areas.

However, the rate of deterioration may be accelerated and major repairs precipitated by failure to take early remedial action. Thus, the accidental puncturing of felt roof coverings may start a chain of events calling for progressively more expensive treatments the longer repair work is delayed, e.g.:

(*a*) Patch roof coverings.
(*b*) Patch roof coverings and make good internal plaster and decorations.
(*c*) Renew damaged timbers and area of felt roofing, make good internal plaster and redecorate room.
(*d*) Renew roof complete including ceiling plaster and redecorate room.

The actual time scale of the events is a matter for experienced judgment based on a knowledge of the prevailing weather conditions and the quality of the construction. It is remarkable that although treatment at stage (*a*) is clearly the cheapest, in many cases the condition of the building is allowed to deteriorate to such an extent that stage (*d*) renewal becomes necessary. Also at this stage there will be the further costs arising from interference with user activities.

It is clear, therefore, that, except where there is a dramatic failure resulting from exceptional weather or other conditions, there is discretion as to the timing of major repairs. However, the deterioration patterns and cost consequences are so ill-defined that decisions are largely intuitive, tempered by availability of finance.

Where the property consists of a number of similar self-contained units, e.g. houses, the increased frequency of repairs to a particular element or component might suggest the comprehensive renewal of all the elements or components of that type throughout the estate. There would usually be some element of improvement in this type of work, in that the object would be not only to reduce the incidence of future maintenance, but also to provide a more up-to-date version of the component to give increased user satisfaction. Thus the timing of the work would depend partly on the difference in cost between bulk replacement of the components and piecemeal replacement of individual components over a period of years, and partly on the degree of obsolescence of the components and acceptability to the user.

Annual programmes

The object of annual programming is to provide a more accurate assessment of the amount of work to be carried out during the forthcoming year and to form a basis for the financial budget. The major considerations would be:

1. Timing the work in relation to the needs of the organisation so as to avoid interference with the basic user activities and in phase with the overall cash flow pattern.
2. Providing a uniform and continuous flow of work for all trades in the direct labour force so as to avoid, on the one hand, slack periods due to insufficiency of work and, on the other hand, the need for men to work overtime at enhanced rates of pay to deal with periods of excessive demand.
3. Fixing an appropriate time scale for the preparation of contract documents and tendering procedures where work is let to outside contractors and for the advance purchase of materials where the work is to be carried out by direct labour.
4. Apportioning the amount included in the budget to specific jobs or areas of work for control purposes.

The annual programme would be built up from:

1. Individual items of painting and repairs brought forward from the long-term plan after a check inspection to ensure that the work is in fact necessary.

2. Individual items of work disclosed by the annual inspection as being necessary to carry out within the next year. The possibility that such work would be needed should have been anticipated when drawing up the long-term programme and a contingency sum included based on past experience.
3. Individual items of work proposed by users at the time of carrying out the inspection. Prior to inspecting premises the user should be asked to state his requirements which, if authorised, would be included in the annual programme.
4. An allowance for work which it is anticipated will be requested by users during the accounting period, but which is not capable of precise definition at the time of the inspection.
5. An allowance for routine day-to-day maintenance based on past records.

The next step is to estimate the probable costs of the individual items of work and the bulk items for routine and emergency work; then to decide, according to the nature and timing of the jobs, whether they are better dealt with by direct labour or by contract. If by the former, the costs should be broken down into labour and material and the labour costs subdivided according to trades.

Table 4.1 shows how it is possible to ascertain the total annual expenditure by trades and the equivalent number of men required in each trade. However, it is unlikely that the number of men so obtained will accord precisely with the number actually employed. Some adjustment will almost certainly be necessary to match the workload to the labour force. This may be achieved by:

1. Adjusting the size or composition of the direct labour force. There are benefits to be obtained from maintaining a stable work force in that the men become increasingly familiar with the needs of the organisation and of one another's methods of working. The result is a better working relationship not only between the maintenance organisation and the users, but also between members of the maintenance team. Also, the continuing association with the upkeep of a particular building or group of buildings tends to develop a greater sense of responsibility, in that shoddy work will only

Table 4.1. Estimated annual expenditure

| Job | Total estimated cost £ | Contract £ | Materials £ | Direct labour force | | Labour | | | Direct overhead costs £ |
				Carpenter £	Painter £	Bricklayer £	Plasterer £	Labourer £	
Individual job									
Job 1	1500	1500	—	—	—	—	—	—	—
2	160	—	30	—	100	—	—	—	30
3	800	800	—	—	—	—	—	—	—
4	265	—	30	100	20	—	—	70	45
5	900	—	150	70	80	250	100	100	150
etc.				(Costs of other individual jobs)					
Routine maintenance	6500	1500	900	700	800	600	800	500	700
User requests and emergencies	3950	750	500	500	300	500	600	300	500
Total	37 550	7550	6200	3500	4800	3500	3500	2500	6000

have to be made good at a later date. Frequent adjustments to the labour force in response to fluctuations in the workload should therefore be avoided.

However, over a period of time changes in the use or physical characteristics of the building may influence maintenance requirements. Such changes should be anticipated in the long-term programme and the labour force either be allowed to decline by natural wastage or be built up to the required strength by a progressive recruiting and training policy.

2. Adjusting the workload. The object of planned inspections is to ascertain the extent of 'necessary' work in the context of overall objectives. Any reduction in the workload must, therefore, involve omitting something which was considered necessary and will result in some loss or inconvenience to the user. Whether or not the work can be deferred until the next budget period will depend upon the cost growth of the repair and consequential losses flowing from the lack of repair. Where the repair cost is not dependent on time and the state of disrepair does not interfere with user activities the delay involves no financial penalties. Such jobs can be used to provide flexibility in the workload provided that the balance from year to year is not affected to too great an extent.

3. Varying the amount of work let on contract. This is usually the best means of levelling out seasonal peaks of activity which are beyond the capacity of the direct labour force. However, the peaks should be adjusted so that, as far as possible, they do not coincide with periods when contractors' prices are at their highest level.

Short-term programmes (monthly, weekly and daily)

So far only broad estimates of costs have been produced under the following heads:

1. Total annual cost subdivided into individual jobs, routine work and emergency work.
2. Cost of work to be let on contract.
3. Cost of work to be carried out by direct labour force,

subdivided into labour by trades, direct supervision and materials.

The next stage is to allocate the total workload to the months of the year in which the jobs will be carried out. Where the work is to be let to outside contractors, the commencement and completion dates should be entered on a bar chart which should also indicate the dates when the various precontract processes should be initiated and completed (see Table 4.2).

The phasing of contract work should take into account not only the convenience of the user and the ability of the maintenance staff to prepare the necessary tender documents, but also any seasonal variations in prices for the type of work to be undertaken.

Table 4.2. Programme for contract work

	APL	MAY	JUNE	JULY	AUG	SEPT	OCT	NOV	DEC	JAN	FEB	MAR
Job 1						$--$ $-\to$		$\longrightarrow$				
3	$-\to$	$\longrightarrow$										
etc.												

$---$ Pre-contract preparation
$\longrightarrow$ Contract period.

The detailed programming of work to be carried out by the direct labour force is dependent upon the size and composition of the labour force and the need to provide continuity of work with a minimum amount of travelling between jobs. At the same time, the availability of materials should be considered and, in particular, the delivery periods for materials not held in store. A practical method of producing this type of programme, suggested by Greenaway,[7] is to prepare a separate strip for each job covering the job duration and indicating the number of men in each trade required each week, as in Table 4.3.

The completed strips are placed on a programming board and the relative horizontal position of each strip adjusted so that the vertical totals show an even balance throughout the

Table 4.3. Strip system of work programme

		Job duration (weeks)			
Job No.	Bricklayer	1	–	–	–
Description	Slater	1	–	–	–
Estimated cost	Joiner	4	4	–	–
	Plumber	2	4	2	–
		etc.			

year, after allowing an appropriate labour availability to deal with routine work and emergencies.

The monthly programmes can then be prepared as in Table 4.4. From this, a daily schedule can be produced indicating the number of tradesmen required daily for each job. However, the problem is rather more complex than would appear from the explanation given, in that it is not enough for the total to remain constant but also the mix of trades should be the same if all the men are to be fully occupied.

The success of these methods depends upon an accurate assessment of the labour content of each job and the sequence in which the individual tasks will be carried out. However, it is quite clear that estimates at this early stage must be based on incomplete knowledge of the extent of the work, and in many

Table 4.4. Monthly breakdown for direct labour

	APRIL				MAY
Working days (20)					
	Carpenter	Painter	Bricklayer	Plumber	Labourer
Total availability	40	60	40	40	40
Job 2		10			
4	10	2			10
5	7	8	22	10	12
Routine maintenance	15	30	10	20	10
User requests and emergencies	10	10	5	10	10
TOTAL	42	60	37	40	42
BALANCE	+2	–	−3	–	+2

cases the precise conditions under which the work will be executed are also conjectural. It would seem, therefore, that the programme at this point in time is of necessity provisional and that the economic deployment of the labour force is dependent on the quality of decisions made very much closer to the time of execution. This is partly conceded in the hospital memorandum which recommends that large jobs should be alternated with a number of smaller jobs, in that the latter can be moved up and down the programme more easily to compensate for errors in the timing of the larger jobs.

It is necessary, therefore, to adopt a flexible approach within the framework of the annual and monthly programmes which serve more as financial yardsticks than rigid schedules of dates for the deployment of labour. The defect of programmes of this type is that the inter-relations between jobs are not made explicit. Thus, where a part of a job is delayed by the non-availability of a particular craftsman, it may be possible to avoid re-timing the whole job by merely altering the sequence of the individual tasks. Although network analyses would undoubtedly display these inter-relationships, the number and smallness of the individual jobs and the frequency of changes would make updating extremely time-consuming.

Clearly the day-to-day jobs requested by users cannot be pre-planned with the same precision. However, short-term planning is both desirable and feasible. Where these jobs are to be carried out by direct labour a simple method is to use a loading board consisting of a set of pigeonholes extending over five or six weeks with slots indicating the number of men of each trade at each location. When a work request is received the planner will assess the number of hours for the particular tradesman and, having checked the availibility of the materials, place the order in the relevant pigeonhole for week 1. When all the available hours have been used up he will start on the next week, and so on. Some rescheduling is possible according to priorities and to achieve more economic groupings of jobs. The foreman will be given all the work orders for the forthcoming week and will arrange the daily programmes according to current labour availability.

Planned inspections

Inspections are undertaken for a variety of purposes, including:

1. Preparing a schedule of the facilities to be maintained and their present condition.
2. Detecting deviations from predetermined standards and incipient faults which may result in such deviations developing before the next inspection.
3. Ascertaining the cause of deviations, the extent of remedial work necessary to restore to the required standard and prevent a recurrence of the defect, and the relative urgency of the work.
4. Checking that previous work was done in accordance with the instructions and that the work specified was adequate.

The main advantages resulting from planned inspections are:

1. An up-to-date appreciation of the overall condition of the property and a corresponding improvement in maintenance records.
2. More accurate prediction of maintenance requirements and hence better budgetary control.
3. A greater proportion of the work can be programmed with less reliance on user requests, so permitting the more economic deployment of direct labour or timing of contract work.
4. A reduction in the risk of breakdowns which may interfere with the use of the building and cause financial loss or inconvenience.
5. The carrying out of timely repairs will extend the life of certain elements and components and reduce the risk of damage to adjoining parts of the building.

The complexity of buildings and the great variety of possible defects makes it necessary fully to pre-plan the inspections and to provide comprehensive check lists to ensure that no part of the building is missed. Appropriate criteria should be laid down for each element or component and the type of inspection needed fully defined and documented. This includes details of the form of the inspection, its frequency, the qualifications of the inspector and the feedback procedures. For greater consistency, special proformas should be prepared for each element and sub-element, setting down the essential information required by the inspector with space for him to record the result of the inspection and the recommended remedial work. In addition, it is advantageous to note the cause of any defect and the urgency with which the work should be carried out. Location codes should be given to facilitate the later processing of the information, but excessively long codes are likely to prove unmanageable in practice.

To facilitate retrieval of inspection sheets, the facilities and elements should be grouped according to:

1. Location. This is of particular importance where the items to be inspected are widely dispersed, in order to devise inspection routes which will minimise travelling time.
2. The periodicity of the inspections. Building elements and materials generally deteriorate at a slow and fairly uniform rate and inspection at annual intervals is usually satisfactory. If more frequent inspections are necessary, provisions should be made to bring the relevant item forward automatically at the appropriate time. A simple manual 'bring up' card index similar to that suggested for hospital plant [8] could be used for this purpose.
3. The qualifications of the person who is required to carry out the inspection.

   (*a*) Operatives where testing and rectification are combined in the same instruction, e.g. checking gutters for accumulation of debris and clearing out if necessary. These jobs involve a straightforward visual inspection which is within the competence of the operative and for which the remedial work can be undertaken immediately without the need for detailed instructions. Control inspections would be carried out at longer intervals of time to ensure that the work was being properly carried out.
   (*b*) Technicians with a sound knowledge of building construction and a broad practical experience would be responsible for the general run of inspections. Such persons would normally require special training in the identification of defects, diagnosing the cause and specifying any necessary remedial work.
   (*c*) Specialists for inspections requiring the use of instruments or the interpretation of legal requirements or for insurance purposes. This would apply particularly to

the inspection and servicing of mechanical and electrical equipment, e.g. lifts. Such persons would be responsible for devising the inspection routines, carrying out control inspections and personally viewing serious defects.

Figure 4.4 illustrates the phasing into the overall work programme of work revealed as being necessary by planned inspections and that requested by users.

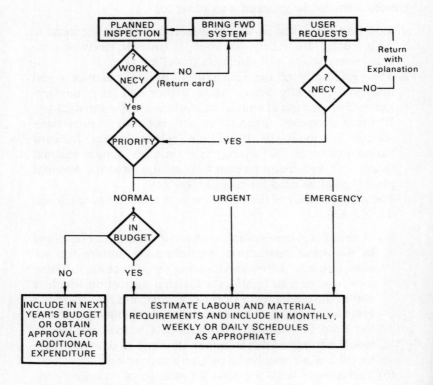

Fig. 4.4. Inspection system

Network analysis

The well-known bar or Gantt chart is a simple graphical method of showing the duration of activities. It is easy to understand and is particularly suitable for communicating short-term site programmes to operatives. However, from a management point of view it suffers from the defect of not showing the relationships between different operations. Critical path method (CPM) is better for the larger and more complex jobs in that the network shows the interdependence of the various operations. Thus in the event of delay it is possible, from an examination of the network, to determine which operations are critical to prompt completion and to concentrate resources on these operations.

A CPM network represents the sequence of operations or activities in a logical manner. An activity is normally regarded as the work done by a man or gang of men but it may also be something which takes only time to perform, such as obtaining approvals. Each activity is represented by means of an arrowed line and starts and finishes at an event (see Fig. 4.5). The circles represent events and are distinguished from activities in that they are points in time and do not consume resources, whether these be labour, materials or time. The events are numbered

Fig. 4.5

and the activities identified by stating the start and finish event numbers. The length and directions of the arrowed lines have no significance and serve only to show the logical dependencies of the activities.

Figure 4.5 indicates that the commencement of Activity 2–3 is logically dependent on the completion of Activity 1–2. If it is assumed that these two activities are 'excavate foundation trench' and 'lay concrete foundations' then clearly the concrete cannot be poured until the trench has been excavated. It may sometimes be possible to break down the job into sections which can be excavated and concreted one after the other. The logical sequence of such an overlapping series of activities can be represented by the use of 'dummy activities', as illustrated in Fig. 4.6. It will be seen that the concreting of the first section, activity 2–3, is dependent on the completion of the excavation of the first section, activity 1–2, and the dummy activity 4–3 indicates that the concreting of the second section, activity 3–6,

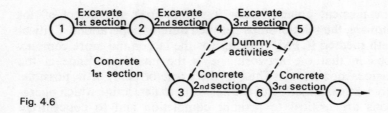

Fig. 4.6

is dependent on the completion of the excavation of the second section, activity 2–4.

Another way in which dummies can be used to preserve the logic of the network is shown in Fig. 4.7, in which activity C can be commenced only on the completion of A and activity D can be commenced only on the completion of both A and B.

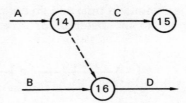

Fig. 4.7

Dummies may also be used to ensure that each activity has a unique reference. Thus if concurrent activities are recorded as in Fig. 4.8 they will have the same start and finish event

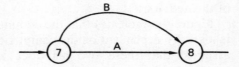

Fig. 4.8

numbers and be indistinguishable when tabulated. In order to avoid this, the dummy activity 8–9 is introduced (see Fig. 4.9).

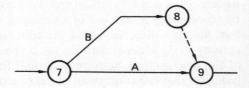

Fig. 4.9

A fault which may appear in a network is that known as 'looping'. Although fairly obvious in the simple example given in Fig. 4.10, it may pass unnoticed in a large and complex

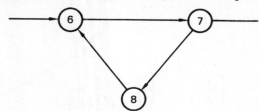

Fig. 4.10

network. On examining the network one discovers that the commencement of activity 7–8 is dependent on the completion of activity 6–7, the commencement of activity 6–7 is dependent on the completion of activity 8–6, and the commencement of activity 8–6 is dependent on the completion of activity 7–8 – i.e. the commencement of activity 7–8 is dependent on its own completion.

For convenience it is usual to call the first event No. 1 and then to number the other events sequentially throughout the network. However, any system of numbering may be adopted provided that no two events have the same number.

The first step in the production of a network is to list the activities in roughly the order in which they will be executed. The level of detail will depend upon the purpose for which the network is being produced and the amount of information available at the time. Then for each activity identify:

(a) the immediately preceding activity;
(b) the immediately following activity;
(c) concurrent activities.

The network can then be drawn as illustrated in Fig. 4.11.

The next stage in the process is to estimate the time requirements or duration of each activity. Any convenient unit of time can be used – hour, working day, week, etc. – according to the magnitude of the project. The durations should be based on recent experience of similar jobs and allow for normal interruptions to the work. In arriving at the duration it may be helpful to adopt the concept used in PERT (Programme

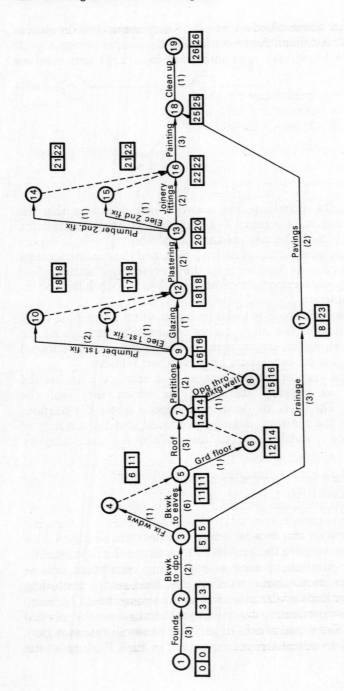

Fig. 4.11. Critical path network for small extension

Evaluation Review Technique) in which the most probable time
(*t*) is obtained from the formula:

$$t = \frac{a + 4m + b}{6}$$

where   *m* = most likely time
         a = shortest anticipated time
         b = longest anticipated time

The durations are then entered in brackets under the appropriate arrowed line.

The next stage is to calculate for each event the earliest time
at which it can be achieved. This is done by making a 'forward
pass' through the network from start to finish, adding up the
durations of the activities leading up to each event. The earliest
starts are recorded in the left-hand box under each event. Thus,
assuming that event 1 starts at time zero, the earliest time that
event 2 can be achieved is day 3, event 3 on day 5, event 5 on
day 11, and so on.

The next computation is to establish the latest time for each
event. This is the latest time by which an event must be achieved if there is to be no delay in meeting the final completion
date. This is obtained by making a 'backward pass' through the
network and is simply a reversal of the method of calculating
the earliest event times. Thus, working back from event 19 to
event 18, there is only one activity emanating from event 18
and as the duration of this has been estimated as one day it is
clear that event 18 must be achieved by day 25 at the latest if
the project is to be completed on day 26. Similarly event 16
must be achieved by day 22, event 17 by day 23, and so on.
The latest times are recorded in the right-hand box under the
appropriate events.

It will be noted that certain events have the same time for the
earliest start and the latest start. These events lie on the 'critical
path' which is the longest route through the network. Any
delay in completing the activities on this path will extend the
overall project time unless compensatory reductions can be
achieved in the durations of subsequent activities. Activities not
on the critical path will have some spare time or 'float'. Initially
activities are planned to start at the earliest time, but to provide
a more uniform use of resources throughout the contract period it may be advantageous to delay the start of some of the

non-critical activities. This is called 'resource smoothing' and for this purpose the network is converted into a bar chart as illustrated in Fig. 4.12, so that the total labour requirements for each day or week can be calculated.

Probably the main advantage of critical path methods is that they demand the logical analysis of the proposed work and provide a means of identifying the particular operations on which resources should be concentrated if for any reason the progress of the work is delayed. Although in this case the method has been applied to the site works it could equally well have been applied to the planning and management processes

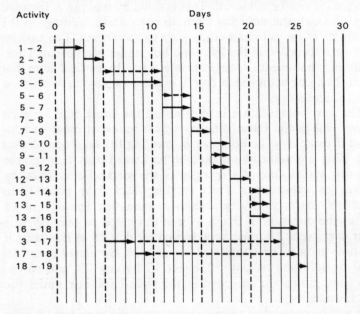

Fig. 4.12. The dashed lines indicate the 'float' or amount of time by which non-critical activities can be delayed without affecting the overall project period (see Fig. 4.11)

and could provide an effective means of control where deadlines have to be met.

An alternative method of producing a network is the 'activity-on-the-node' system or, as it is sometimes called, 'precedence' diagram. In this system the activity is represented at the node which is the equivalent of the event in the critical path method. It is generally thought to be easier to understand than the conventional arrow diagram and is frequently used as a preliminary form of analysis. Figure 4.13 illustrates the

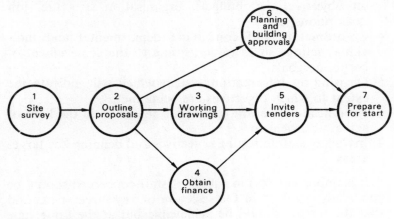

Fig. 4.13. 'Activity-on-the-node' system

application of the system to a hypothetical job programme. The activities are written in abbreviated form in the circles (or rectangles if preferred) and each one is given a unique number. The duration times may be entered subsequently under the activity description.

Management by objectives (MBO)

This is a discipline which attempts to define individual responsibilities by setting objectives which have to be achieved within key target areas (KTAs). The object is to integrate an organisations needs by involving management and subordinates in the process of deciding what to do, how to do it and when to do it. The method provides a rational basis for determining corporate and/or departmental goals and how to achieve them. The basic elements of the method are:

1. Planning of corporate objectives relating to every area of performance which affects the organisation, i.e. long-term planning.
2. Planning corporate, departmental and individual ob-

jectives which logically derive from the long-term plan, i.e. short-term planning.
3. Aligning other constraints on managerial action with the set objectives, i.e. budgets, organisation structure, job descriptions, etc.
4. Co-ordinating and controlling departmental and individual actions so as to contribute to the achievement of corporate goals.
5. Devising an information system which will indicate the extent to which objectives are being met.
6. Periodically reviewing goals and objectives in the light of actual achievement.
7. Involving staff in setting objectives and defining key target areas.

It is important that members of staff concerned should be personally involved in the selection of key target areas and that the targets should be achievable but at the same time present a challenge. Sherwin[9] has suggested that objectives fall into two distinct categories:

1. Functional performance objectives, i.e. those which maintain predetermined standards and are repetitive such as:

    (a) Routine maintenance.
    (b) Painting externally every five years.

2. Change objectives, i.e. those that require some distinct cognitive and creative effort and are usually accompanied by some changes in policy, procedures, personnel or investment such as:

    (a) Reducing backlog of work.
    (b) Increasing speed of response to tenants' requests.
    (c) Improving productivity of labour force by 25 per cent.

Thus the organisation for change objectives is essentially a temporary one, usually requiring co-operation from different groups, but on achievement the change objectives become functional performance objectives.

It is important to have regular meetings with participants to discuss progress and possible modifications made necessary by unforeseen difficulties. There should be an annual report giving information on such matters as any revisions to the objectives, reasons for failing to meet objectives and details of the steps taken to ensure that the same problems will not occur again. These evaluations are very useful in the long-range planning process in making the department's future objectives more realistic.

Maintenance audits

Every organisation should have some means of checking its management efficiency. Unfortunately in the case of maintenance there is no single measure that is appropriate and it is necessary to analyse the answers to a series of questions. Like an accountant's audit the essential feature of a maintenance audit is that it seeks to evaluate the system of internal control, to determine whether or not it is being operated efficiently. Of course, it presupposes that there is a definable maintenance management system to audit.

The audit can be broken down into the following areas:

Technical

What is the general condition of the building(s)?
Is the standard of maintenance appropriate?
Have the defects been correctly diagnosed?
Is the remedial work satisfactory?
Are the users satisfied with the quality of the work and the speed of response?

Management

Is there a maintenance policy?
What is the organisational structure?
Are the operating procedures clearly defined?
What records are kept and are they up-do-date?
What percentage of the work is planned?
What is the percentage of emergencies?
What is the average backlog?
Is there an efficient work order system?
What controls are exercised over time, cost and quality?
What is the involvement, if any, of the design of new buildings?

Financial

What is the cost per unit, e.g. per dwelling, per job, etc.?
Is the cost within authorised expenditure limits?
What is the return on capital for direct labour?
What is the ratio between total costs and administrative costs?
What are the administrative costs per dwelling, per work order, etc.?

Direct labour force

What is the gross output per man?
What is the value of materials used per trade?
How many jobs are completed within a given period?
What are the man-hours per job in each trade?
How are the operatives motivated?
What is the percentage of non-productive time?

Contractors

How are contractors selected?
What is the basis of payments?
What contractual controls are there over time, cost and quality?
What is the average difference between final account and tender?
What is the average delay in completion?
Is the quality of the work satisfactory?

It will be apparent that operating such a system of control calls for the full cooperation of all members of the maintenance staff, and to achieve this clear guidelines and procedures should be laid down in a Maintenance Practice Manual, as outlined in Fig. 4.14.

Maintenance practice manual

1. General
   Objectives and policy
   Staff duties and responsibilities
   Maintenance categories and priorities

2. Planning and Control
   Approach to planning maintenance
   Work order system
   Estimating methods
   Performance checks

3. Inspections
   Type of inspection
   Period
   Check list and report

4. Budgetary Control
   Preparation of budget
   Cost centres
   Variance reports

5. Execution of Work
   Direct labour force
      Planning and control system
      Bonus payments
      Stores, plant and transport
   Contractors
      Tendering procedures
      Types of contracts and documents
      Supervision and contract administration
      Claims and payment

6. Information System
   Records
   Paperwork/computer system
   Analysis of information
   Feedback and reports

7. Health and Safety Measures

8. Staff Recruitment and Training

9. Tenant/User Relations

Appendices
   Standard forms
   Flow charts
   References

Fig. 4.14

References

1 Robertson, J. A. (1969) 'The planned maintenance of buildings and structures.' The Institution of Civil Engineers, Proceedings Paper 7184S.
2 Harper, F. C. (1968) 'Maintenance of buildings.' Bath: Conference on Technology of Maintenance.
3 The Tavistock Institute (1966) *Interdependence and Uncertainty – Study of the Building Industry*. London: Tavistock Publications.
4 Reading University (1982) 'Housing authorities tendering and contract procedures' (unpublished).
5 Speight, B. A. (1969) 'Formulating maintenance policy.' London: Conference on Building Maintenance.
6 Luke, R. (1963) 'Practical aspects of planned building maintenance.' London: International Maintenance Conference.
7 Greenaway, L. R. 'Maintenance of buildings – potential savings greater through improved control.' *Contract Journal*, 3 Dec. 1964, pp. 659–60.
8 Department of Health and Social Security. *Planned Preventive Maintenance: a System for Engineering Plant and Services*. Hospital Technical Memorandum No. 13. HMSO, 1968.
9 Sherwin, T. S. (May/June 1976) 'Management of objectives.' *Harvard Business Review*.
10 Skinner, N. P. 'The matching of planning systems to the maintenance work that is actually done.' Edinburgh: CIB/W70 Seminar, March 1983.

CHAPTER FIVE

Cost control

Estimating accuracy

The value of planning depends to a large extent upon the accuracy with which future performance can be estimated. However, the degree of accuracy should be consistent with the purpose for which the estimate is to be used and the time-scale of the predictions. An excessively detailed approach can sometimes result in unjustifiable costs and may detract staff from other duties. As stated by a spokesman for a large organisation, management and supervisory staff are liable to become predominantly involved in this process to the detriment of the essential work of management and supervision of the current programme. He concluded that it was better to be less meticulous in estimating and to concentrate the maximum effort on management, supervision and control of expenditure. However, the preparation of estimates will create a need to plan jobs in advance and will thus stimulate more effective management.

The accuracy of the estimate will depend upon the amount of information available on:

the nature and extent of the work
the conditions under which the work will be executed
the mode of execution
the costs of employing labour
the prices of materials.

Clearly, the longer the time-scale of the estimate, the more difficult it will be to predict the above factors and the more uncertain the estimate will be.

Long-term estimates

Long-term estimates may extend over a number of years and be required for a variety of purposes, including financial planning on a national scale for public buildings such as hospitals or schools or long-term maintenance programmes for individual buildings. The characteristic feature is that the precise nature of the individual items of work is not known and the estimate must be based on the average cost of maintenance related to some parameter of the building or buildings in question. Methods include the following.

Financial criteria

Various ways have been suggested of relating maintenance costs to other costs or receipts. These include expressing maintenance costs as a percentage of:

1. Construction costs. This should be the current replacement value in that the initial cost of construction would give a misleading picture in this context. Unfortunately, replacement value is often merely a 'book' figure and does not accurately represent the actual cost of rebuilding. Also the percentage will vary over a period of time according to the age of the building and the different rates of increase of maintenance costs and new construction costs. There will also be a between-buildings variation according to the type of construction, e.g. higher initial costs may have been incurred to avoid maintenance and theoretically this should give an inverse relationship.
2. Production costs. These are determined by factors quite different from those which generate a need for maintenance, e.g. product mix, number of shifts, efficiency of production equipment, etc. Indeed, it is possible that a reduction in production costs by greater mechanisation may increase the maintenance costs.

3. Occupation costs. Again there may be an inverse relationship between maintenance and certain other occupation costs: e.g. fabric maintenance may, by improving the insulating properties, reduce heating costs.
4. Profitability. Maintenance is only one of a great many factors which determine profits; others, such as market conditions, level of prosperity, demand for goods produced or sold, advertising, etc., have a much more direct effect. It should be borne in mind that while a fall-off in sales may be temporary, a reduction in maintenance can have costly long-term consequences.

Relating maintenance costs to the above does not take into account the particularity either of maintenance costs or of the other expenditures. It therefore provides only a crude guide where there is some similarity of use and construction.

Costs per unit of accommodation

The unit would be related to the number of people using the building, e.g. pupils for schools, or to functional units, e.g. general wards in hospitals. It is useful for rapid assessments of the probable maintenance expenditure on a national or regional basis for buildings of the same use category. The disadvantage is that the unit costs are not capable of precise adjustment according to the particular problems encountered in individual buildings.

Costs per unit of floor area (m²)

Again, this is a quick and simple method of obtaining a rough estimate of maintenance costs. Different use areas will attract different levels of maintenance expenditure and should be kept separate. The unit costs are obtained from past records but require adjustment for age, constructional form, etc., if applied to other buildings.

Costs per unit of volume (m³)

As in the floor area method, the different use areas should be treated separately. However, both methods suffer from the defect that there are other factors which influence the level of expenditure. In a study of hospital buildings it was found that the most significant factors in addition to volume were the numbers of buildings and of occupied beds. A formula for estimating maintenance costs was derived by regression analysis in the following form:

Maintenance costs $= x$ (volume) $+ y$ (no. of buildings)
$+ z$ (no. of occupied beds)

The coefficients relate to one year only and require updating according to the average maintenance costs per unit volume for all hospitals in the year under consideration.

Costs per building element

The assumption is that an analysis of past maintenance costs according to building elements, i.e. external walls, floors, roof, etc., will reveal a pattern of expenditure which will suggest reasonable amounts to spend on each element. The costs are related either to the total floor area of the building or to the area of the particular element. However, there are clearly no absolute or ideal values, and in many cases the elemental costs will be cyclic. A study of expenditure over a period of years will indicate average costs and also the periods for planned maintenance. In addition, a sharp increase in the repair costs for an element would prompt an investigation to discover the reason – whether or not renewal is desirable or if the element is being subjected to abnormal use.

Costs per functional system

This method attempts to relate maintenance costs more closely to the needs of the user and thus to the benefits which flow from different levels of expenditure. It thus provides a more realistic basis for policy decisions than an elemental breakdown, in that the cost of maintaining each facet of the environment is made explicit. The method of subdividing into functional systems is described in Chapter 6.

Medium- and short-term estimates

The methods so far discussed are attempts to determine at an early stage an 'ideal' amount based on historic costs and related to standards which were considered acceptable. Later, for programming, scheduling and controlling the execution of the work, more accurate estimates are needed based on the actual work which has to be undertaken. The methods adopted include the following.

Analysis

Small jobs can be broken down directly into their labour and material content on the basis of past records of similar jobs. These are usually single-trade jobs which consist of a straightforward uninterrupted work sequence. Having established the labour hours consistent with the assumed job method and working conditions and calculated the material quantities, these are priced out at the rates which it is predicted will apply at the time of execution. Larger jobs must first be broken down into a series of separate items representing discrete parts of the whole. The items usually include both labour and materials and may refer to a single trade operation or a self-contained multi-trade operation. The items are then analysed to determine the labour and material content and priced as for small jobs. For example small jobs may be analysed directly into labour and material as follows:

Hack out broken pane of glass size 450 × 600 mm and reglaze with 3 mm OQ glass in putty to wood casement.

	Item
Labour hacking out broken pane, preparing rebates and reglazing, 1 hr @ £4* per hr	4·00
Pane of 3 mm OQ glass size 450 × 600 mm	2·50
Sundries – sprigs, putty, etc.	0·50
	7·00
Overheads and profit +20%	1·40
	£8·40

* All-in rate but excluding travelling time, etc.

Larger and more complex jobs may require to be broken down into the constituent items as follows:

Cut opening size 790 × 2072 mm through 100 mm block partition, insert 100 × 150 mm precast concrete lintel over, square up jambs and make good plaster and skirting both sides, extend vinyl floor tiles through opening and provide and hang new flush door 40 mm thick × 726 × 2040 mm to 32 mm thick linings with 19 × 75 mm architrave both sides, provide and fix mortice latch and furniture and knot, prime, stop and paint one undercoat and one finishing coat on new woodwork.

			Item
Cut opening through block-work and clear away debris	2 m²	£8	16·00
Precast concrete lintel and wedge and pin up over	1·25 m	£12	15·00
Square up jambs	4 m	£2	8·00
Make good plaster up to lining	10 m	£2	20·00
Ditto vinyl tiles through opening	1 m	£6	6·00
Remove skirting both sides, cut and fit to linings and refix	Item		10·00
Flush door including hanging and hinges	Item		30·00
32 × 125 mm door linings	5m	£2·50	12·50
19 × 75 mm architrave	10 m	£1·50	15·00
Mortice latch and furniture	Item		15·00
KPS and 2 coats paint	5 m²	£2	10·00
			157·50
Overheads and profit +20%			31·50
			£189·00

It will be seen that the analyses consist of three main elements – labour, materials and overheads and profit.

1. Labour

Labour times are obtained from records of past jobs, time studies or activity sampling surveys or, if the information is not available within the organisation, from one of the published price books such as Spon's or the **BMCIS Building Maintenance Price Book.** Great care should be taken when using published prices in that labour times for maintenance operations vary between very wide limits. The main factors that should be taken into account are:

(*a*) Travelling time. The cost of a small maintenance job will depend to a large extent upon whether it is a small isolated job remote from the depot or one of a series of similar jobs which can be carried out at the same location.

(*b*) Access to the building. A good deal of time may be wasted if repeated visits have to be made before gaining access to the building.

(*c*) Working conditions. Apparently similar jobs can present very different problems to the operative according to the height at or position in which the work has to be carried out.

(*d*) Degree of repetition. Clearly a familiar frequently occurring job will take less time than one which the operative has not done many times before.

(*e*) Timing of work. If the work has to be carried out outside normal working hours allowance should be made for overtime rates.

(*f*) Pre-inspection. In some cases it may be necessary to pay a preliminary visit to the site to ascertain what has to be done.

(*g*) Preparing and cleaning up on completion. The labour constants given in price books assume that the operative can commence the work immediately and that when he has completed the job he moves straight on to the next job. A casual observation of the behaviour of workmen will reveal that this does not happen in practice. In an occupied building it may be necessary to remove carpets and furniture before starting, the progress of the work may be affected by the need to keep down dust and noise and on completion the premises must be left clean and tidy.

(*h*) Labour on-costs.

The all-in hourly rates may be calculated in accordance with the Institute of Building Code of Estimating Practice.[1] The additions to be made at the basic wage rates for building operatives are:

Bonus
Overtime
Cost of living supplement
Sick pay
Tool money
Holiday credits
National insurance and pensions
CITB Levy
Severance pay
Employer's liability and third-party insurance
Trade supervision
Joint Board supplement

Currently the above allowances add about 66 per cent to the basic wage rate.

2. Materials
Allowance should be made for:
(a) Wastage and breakages.
(b) Quantity – it should be remembered that materials purchased in small quantities will cost much more than the price given in published price books.
(c) Transport of materials to site including unloading and possibly storing prior to use.

3. Overheads and profit
The overhead costs are the costs of administering the organisation and include:

(a) Head-office costs.
(b) Stores and yard.
(c) Supervisory staff.
(d) Small plant, hand tools, ladders, etc. (the cost of large items is better included in the rate for the particular item of work for which it is used).
(e) Welfare and safety provisions.
(f) Profit would be included where the object of the estimate is to forecast the amount of the lowest tender from an outside contractor. It would not be included where the work is to be carried out by direct labour unless the direct labour organisation is treated as a trading organisation when the profit element would represent what is regarded as an acceptable return on capital.

The disadvantages of analytical estimating for maintenance are:

1. Lack of standardisation of job descriptions make precise identification difficult.
2. Job times are rarely available in sufficient detail to permit adjustments to be made for slightly different conditions.
3. The estimated times are usually averages which are of little value for programming and financial control.
4. The recorded times may not distinguish between productive and non-productive time, with the result that the reasons for wide differences in the times of apparently similar jobs are not made clear.
5. The recorded times are not necessarily the optimum times and do not indicate the quality of the management.

Judgment

In many cases historic cost data are so scanty that estimates are based solely on 'experienced' judgment. This applies particularly where the foreman is responsible for estimating and usually takes the form of a total cost based on the number of men required and the time they will take to complete. Although this is quick and, in view of the foreman's familiarity with the work and ability of his men, reasonably accurate, it has certain disadvantages: principally that consistency cannot be proved and insufficiently detailed information is provided for feedback purposes. The main justification is that it avoids duplication of effort in that where there is a separate estimator the foreman must still analyse the job for labour deployment and requisitioning materials. However, it distracts the foreman from his main task of supervising the work; also, for financial control an independent estimate is to be preferred.

Slotting

Small non-repetitive jobs may be dealt with by a technique known as 'slotting' or 'bracketing' – that is, jobs are classified within time brackets by reference to typical common jobs. For a suitable base period the total number of jobs and the average time per job are recorded in each of the following groups:

Jobs taking from   0·1 to   8·0 hours
8·1 to 16·0
16·1 to 32·0
32·1 to 48·0
48·1 to 96·0

From the frequency and average time of jobs in each group the total time and labour costs for a period can be calculated. The cost of materials can be expressed as a percentage of the total labour cost. The method is appropriate for estimating the budget allowance for contingency maintenance for which the precise nature of work requested by users cannot be forecast, but which nevertheless may follow a regular statistical pattern.

For very small day-to-day routine jobs, an average cost per week can be derived from past records for the different categories of work.

Approach to estimating

The approach to medium- and short-term estimates will vary according to whether the work is scheduled to be carried out by direct labour or by contract.

Direct labour

In addition to forecasting the organisation's financial commitments, estimating has important functions in relation to the control of directly employed labour:

1. The total estimated labour hours will indicate whether or not the existing labour force has the capacity to do the work and the nature of changes which should be made to its structure and composition.
2. The estimated job times will form a basis for programming and scheduling labour resources.
3. The job times also provide a yardstick against which to

measure actual performance and provide a basis for bonus targets.

4. The aggregate job times permit an assessment to be made of the backlog of work which exists at any time so that the response time can be adjusted if it is unacceptably long or uneconomically short.

The emphasis is therefore on estimating realistic times for the jobs based on the known capabilities of the men employed.

Contract

Where work is let to an outside contractor the primary object of the estimate is to predict the amount of the lowest tender or, where it is let on a cost reimbursement basis, the final cost. Unfortunately, information from past contract work is apt to be very much less detailed than that from work executed by direct labour. The main reasons for this are:

1. The method of presenting information to the contractor for tendering purposes. Often the contractor is given only a very general idea as to what is required and must use his own judgment as to the scope of the work. Alternatively, he may be presented with a brief schedule of composite items couched in terms which seek to cover every possible contingency. In both cases the result is a lump sum undivided tender which allows for additional work which the contractor assumes will be necessary, but which is not identified. Even where more detailed schedules are produced the items are usually unique to the particular job and, in the absence of further subdivision, cannot be applied to other jobs.

   However, whereas whole jobs rarely repeat themselves in precisely the same form, they are made up of operational tasks, i.e. short single-trade work sequences, which do tend to be repetitive. It would therefore be helpful if the estimates were related to standardised tasks for which average times could be built up over a period of time.

2. The contractual arrangements. Maintenance work is frequently carried out on a cost-plus basis and the final account shows only the total labour hours worked and

materials used. For jobs of any size it is impossible to allocate the costs to particular sections of the work.
3. Job method and sequence of operations. Unless the user of the building demands a particular job method, it is usually at the contractor's discretion and will not be known at the time of preparing the estimate.
4. Tendering climate. In addition to the basic costs of labour and material, it is important to consider the keenness or otherwise of the contractors to secure the contract. This will depend upon the nature and timing of the work and the degree of risk involved. It will also depend in part on extraneous factors such as the volume of other, possibly more attractive, work available in the locality at the time of tendering. Thus the estimate is not in any sense absolute and any change in the programmed dates may render it inappropriate.

Financial control

The main instrument of financial control is the budget. This is a predetermined statement of management policy during a given period and it provides standards for comparison with results achieved. It has been described as a means of equating available financial resources to planned expenditure. This definition presupposes that maintenance will be planned some time in advance of execution and that resources are limited. Most budgets are established on an annual basis to coincide with the conventional accounting periods, but for maintenance a longer-term budget is desirable, even though some of the items must of necessity be provisional.

The maintenance costs form only a part, perhaps quite a small part, of total operating costs. Indeed, for manufacturing industries the cost of maintaining buildings is usually less than $\frac{1}{2}$ per cent of turnover. In view of the comparative smallness of the expenditure, there is a danger that insufficient attention will be paid to determining the optimum amount to spend on maintenance. This can result in either inadequate funds being set aside for maintenance or an attitude of indifference. The effects of any resulting under-maintenance may not be immediately discernible, but may lead to extensive remedial work being necessary in subsequent budget periods. The long-term plan displays a better picture of the ultimate cost of neglect.

It has been recommended that the budgetary period should be related to the rate of deterioration of a significant element of the building stock. Thus, where circumstances dictate a five-yearly cycle for external painting, the budget period should be based on this. The annual programme may then be seen as a part of a continuing series of work matched to the organisation's cash flow. Shorter-term budgets – say at monthly intervals – are necessary to accommodate unforeseen work or changed conditions.

Budget preparation

The normal procedure is for departments to produce sectional budgets which are considered in the light of the firm's broad policies, amended if necessary and finally integrated into a comprehensive plan. The financial content of budgets and the close links between budgets and accounts usually result in the controlling and co-ordinating function being exercised by someone with accounting skills. Any pruning to bring the estimates into line with available resources is therefore likely to be to those activities which do not have an obvious and immediate impact on production or which are not shown to produce quantifiable benefits.

In a study carried out by Bath University it was found that it was rare that anything more than a 'guestimate' was made at the budget stage and that where cost estimating was attempted it was elementary. The general conclusion reached was that technical assessments were not sufficiently authoritative to withstand pressures of demands from other users of resources. A similar conclusion is expressed in the Woodbine Parish Report in relation to hospital maintenance organisations. Clearly, if the maintenance budget is presented as a total sum, possibly with large contingency amounts and no indication of the benefits accruing from carrying out the work or of the penalties which will be suffered if the work is deferred, it is possible that funds will be diverted to other apparently more profitable activities.

In such circumstances, much will depend upon the firm's attitude to maintenance and on the personality of the maintenance manager and his skill in presenting his case. Often one finds that the maintenance organisation has evolved over a period of years in response to the expansion of the firm's major activities but that the original organisational structure and operating methods remain unchanged in spite of the new demands. The resulting great diversity in the approach to maintenance budgeting with different methods of classifying and recording costs makes interfirm comparisons in this field very difficult. In the majority of cases, maintenance and other occupation costs are fragmented on a functional basis and control is exercised by different people. Thus the interactions between different elements of occupation costs are not made explicit and it is not possible in this situation to devise a coherent maintenance policy.

Subdivision of budget

The benefits of operating a system of budgets are not maximised unless procedures for budgetary control are instituted. The budget should therefore be subdivided in such a way as to permit both financial and performance control. Thus, in addition to stating estimated costs it should also detail the plan of activities and lay down standards against which actual performance can be measured and deviations identified for investigation. It is important that the budget should be properly and clearly presented in order to convince upper management that the proposals have been properly thought out and are both feasible and necessary for the achievement of the overall objectives of the organisation. The maintenance budget should therefore be comprehensive and show clearly the relevance of the proposed expenditure to user needs and the long-term overall policy of the organisation. In particular it should make explicit the consequences of neglecting to carry out essential work.

The following is a typical form of presentation for the maintenance budget:

1. Title
Maintenance budget (period).

2. Statement of policy

An outline of maintenance policy and the relevance of this policy to the broader objectives of the organisation. In particular any changes in maintenance policy should be stated, giving reasons and anticipated benefits.

3. Breakdown of proposed expenditure

The expenditure should be subdivided in various ways to indicate the nature of proposed work and the associated costs. Possible classifications are:

(a) By type of costs.
(b) By type of work.
(c) By location.
(d) By method of execution:

  (i) Direct labour − break down into manhours per trade, materials and plant.
  (ii) Contract − indicate types of contract to be adopted.

4. Discretionary items

This section would cover major works which are not strictly maintenance but which are thought desirable in that they would permit the building to be used more efficiently. Such works would include:

(a) Replacements of major components.
(b) Alterations and extensions to existing building.

An analysis should be given of the anticipated costs and benefits so that the merits of the proposals can be judged.

5. Cash flow

A programme should be included indicating the timing of the proposed work and showing the cash flow pattern over the budget period.

6. Supervisory and clerical staff

Information should be given on any proposed changes in the structure of the maintenance organisation together with any re-allocation of staff duties. If necessary this would be accompanied by requests for the upgrading of existing staff or the appointment of new staff.

7. Direct labour force

Details should be given of the following where appropriate:

(a) Number of men employed and anticipated wage bill.
(b) Reasons for any change in the number as compared with the previous period.
(c) Purchase of additional plant with reasons.
(d) Introduction of incentive scheme.

8. Appendices

(a) Outline of annual inspection report.
(b) Outline of long-term programme to show timing of major works.
(c) Summary.

Types of cost

Costs may be classified in a variety of ways according to the accounting procedures of the organisation. The basic division is into capital for the provision of new assets and revenue for maintenance and running costs. There are taxation implications in that whereas expenses of a revenue nature can be set against taxable profits, those of a capital nature cannot. Jarman[2] has suggested that, where new capital assets are considered, the financial approval should cover not only the initial cost but also the setting aside of monies to ensure that the necessary funds will be available at specific times for replacement and routine maintenance. However, such reserves are not allowable deductions for tax purposes.

Ray[3] has distinguished between the different types of cost as follows:

1. Committed costs which represent the after-effect of irreversible decisions taken in the past. Such costs have a mandatory character and include those which are incurred in complying with the terms of a lease or statutory requirements. There is, of course, some element of discretion if the firm is prepared to risk a civil action for damages or a criminal prosecution

for disregard of statutory provisions. Other committed costs would be those flowing from specific decisions taken at the design stage, say, to use short-lived finishing materials on the assumption that they will be renewed at predetermined intervals. Again, the costs may be avoided if the occupier is prepared to accept some loss of amenity or a degraded appearance.

2. Variable or engineered costs which can be directly related to the volume of the primary activity of the organisation, e.g. products manufactured or sales. The difficulty lies in calculating the right amount for a given level of activity. Clearly, where this can be done it provides a simple means of arriving at the permissible level of expenditure on maintenance. However, the rate of deterioration of many building elements is unaffected by user activities and the state of disrepair of these elements would have to be extreme to influence markedly the efficiency with which the activities are carried out. Where a system of planned maintenance is in operation, the object would be to carry out repairs well before such extreme conditions were reached.

3. Managed or discretionary costs which require specific decisions in each budget period. This method is criticised on the grounds that there is no scientific method which can determine the right amount which must, therefore, be a matter of judgment. As managed costs represent an 'agreed' amount to spend and not necessarily the 'correct' amount, actual costs below the budget are no reflection of efficiency.

As far as efficiency is concerned, there is clearly a difference between the efficient use of resources in the context of the organisation as a whole and the efficiency of the maintenance department. If the agreed amount includes unnecessary work, then no matter how efficiently the work is executed, it will represent waste. It is thus a matter of identifying necessary work and this demands the exercise of judgment in assessing the probable rate of deterioration and related cost growth, identifying the cause of the defect and specifying the most effective and economic remedial measures. Another alternative is that although the items of work may be necessary, their estimated cost is extravagantly high. Clearly, in such cases actual costs lower than those budgeted would merely reflect poor estimating.

Types of work

Maintenance falls naturally into two main categories:

1. Minor items which, although individually of insignificant cost, in total account for a significant proportion of the workload. These are distinguished by the fact that although the timing is uncertain, the total tends to remain constant over a period of time. The inclusion of a lump sum in the budget is open to criticism on the grounds that while this would give financial control, it does not provide a yardstick for performance control. However, the precise identification and costing of each individual item would probably be more costly than any savings it would achieve.

   A compromise would be to subdivide the lump sum according to trades, cost range of jobs and frequency. This would necessitate a critical examination of historic costs over a period of years to determine the pattern of expenditure. Checks could then be made at intervals to ensure that the work ordered under this heading was both necessary and efficiently executed.

   Alternatively, control could be exercised nearer the time of execution by close scrutiny of the work ordered, and the calculation of the work content of each job in standard hours for comparison with actual hours worked. This could be coupled with a work-measurement-based incentive scheme so that the system serves both to motivate and control.

2. Major items which, with the exception of jobs necessitated by abnormal conditions, can be predicted well in advance. These items fall into two further categories:

   (*a*) Cyclic work such as external painting and internal decorating, including pre-painting repairs.
   (*b*) Replacement or renewals to eliminate high cost areas.

   At the same time, consideration would be given to alterations and additions to improve amenities. Although this work is not strictly maintenance, it is usually the responsibility of the maintenance organisation in that it can be carried out at the same time as associated repairs and by the same operatives. Whether or not renewal is worthwhile in purely economic terms can be easily calculated by comparing the combined annual running costs and amortised initial costs of the replacement with the present running costs. The following information is required for the assessment:

(a) The present running or repair costs and, where the cost increases with time, the predicted rate of growth based on an analysis of past costs and judgment.

(b) The costs of replacement including where appropriate an allowance for disruption and loss of service during the period of carrying out the work.

(c) The anticipated life of the replacement over which the cost can be amortised.

(d) The running or repair costs of the replacement and, if this is not constant from year to year, the estimated rate of increase.

(e) Any quantifiable benefits which will be gained from the replacement. Where the benefit is increased and user satisfaction arises, for instance, from the better appearance of the replacement, the benefits can only be evaluated subjectively unless some closely associated cost can be used as an indicator.

The techniques for rationalising repair/replace decisions were described in Chapter 1. When applying these techniques account should be taken of the rate of growth of the repair costs in relation to both the original element and its replacement. Also, there will come a time when further repairs are not possible and the component must of necessity be replaced. The object is to determine the most economic and convenient time to carry out the work.

Variance reporting    FEED BACK

A system of reporting at regular intervals should be instituted so that any difference between actual costs (including committed costs) and those forecast can be analysed, the reasons ascertained and, if necessary, corrective action taken. Staveley[4] suggests that control should be exercised from a study of monthly statements of expenditure against budget allocation and recommends that:

1. If overrunning the budget, either

(*a*) defer non-programmed work for inclusion in the following year's budget, or

(*b*) reduce the scope of programmed work.

2. If under expenditure, bring forward major repairs.

Presumably the non-programmed work referred to is that arising from user requests for which a general allowance was made in the budget. However, it would be difficult to apply these remedies during the latter part of the budget period. In particular, bringing forward major repairs might result in inadequate time being available for the proper planning of the work or in the work having to be carried out at an inconvenient time. Before deciding upon the remedy, the reasons for the variance should be carefully analysed. These may be:

1. Tenders in excess of the estimate due to
   (*a*) Inaccurate estimates.
   (*b*) Unforeseen increases in wage rates or materials prices.
   (*c*) Change in market conditions affecting tendering climate.
   (*d*) Delay in starting the work resulting in tenders being invited at an unfavourable time.

2. Increase in scope of the work due to
   (*a*) Variations during execution period or additional work revealed when surface coverings removed.
   (*b*) Higher proportion of user requests than anticipated.
   (*c*) Exceptional weather conditions or abnormal user.

3. Inefficient organisation and planning of work resulting in
   (*a*) Higher proportion of non-productive time.
   (*b*) Lower output during productive time.
   (*c*) Materials wastage.

It will quite frequently be apparent that an increase in the budget allowance would be more appropriate than deferring the work, in that the cost of not doing essential repairs might well be greater than the cost of carrying out the work during the current budget period. Also, monthly fluctuations may be misleading and cumulative trends are likely to be more informative. Maintenance expenditure will not normally be spread evenly over the year since not only do user requirements often dictate the carrying out of major repairs at particular times, but also it may be possible to take advantage of seasonal variations in contractors' price levels. The monthly divisions of the budget should therefore be related to the work programme and not, as is usual accountancy practice, one-twelfth of the annual amount. Any adjustments to the work programme necessitated by unavoidable delays should be reflected in corresponding amendments to the monthly budget. Account should also be taken of the time-lag between execution of the work and payment, and of the fact that some part of the monies due to a contractor is usually retained for an agreed period after completion as a protection against defective work.

Performance measures

In order to determine the effectiveness of planning and improved working methods, it is necessary to have some means of measuring performance so that before-and-after comparisons can be made. In building maintenance the lack of a standard unit of production and the extremely varied conditions under which the work is carried out preclude precise comparisons. To some extent it is possible to compare the amount of the work done by two men by an examination of their time sheets, but for larger numbers of men the relationship between work done and time taken must be converted into figures which are more readily comparable. In addition, the measure should have an interpretive function in that it should indicate the nature of errors in past decisions and suggest appropriate remedial action. However, in order to achieve this several related measures are needed covering the various factors which, in combination, determine the level of performance. The measures should be easy to calculate and use and in most cases consist of simple ratios. Such ratios must be used with care as they reflect relative rather than absolute values and the same ratio may be produced by completely different circumstances. For this reason the following performance measures have been grouped broadly into those which attempt to measure productivity and those which have an explanatory function.

Measures of productivity

1. Gross output per operative

$$\frac{\text{Total cost of wages, materials and overheads}}{\text{Number of operatives employed}}$$

This gives a crude index of productivity but suffers from the following defects:

(a) The total cost will depend upon the proportions of the different trades and the relative costliness of the materials they use. It is not suitable, therefore, for intertrade comparisons, e.g. electricians use more expensive materials than painters and their output per man expressed in monetary terms is correspondingly greater.

(b) By paying overtime and bonuses the total cost is increased without necessarily involving a higher rate of productivity.

(c) Increases in overhead charges give an apparent increase in productivity. Also, as methods of calculating overhead charges vary so greatly, the figure is not suitable for interfirm comparisons.

(d) In some cases the cost of work to be done is decided in advance by estimates which are political targets rather than measures of what needs to be done.

2. Value of materials used

This may be expressed in terms of the total value of materials over the period under review or the value of materials per operative or gang. Although the method gives some measure of the amount of work done, its usefulness is limited as follows:

(a) The materials used by different trades vary widely in cost and even within a particular trade there will be different proportions of cheap and expensive materials and of labour and materials costs in different periods.

(b) The greater the wastage of materials, the higher the apparent output.

3. Orders executed

This method is only viable where there is a high degree

of repetition. It otherwise suffers from the following defects:

(a) The number of work orders completed in the period reflect only the size and complexity of the jobs dealt with. For the smaller, more repetitive jobs the orders may be banded according to average times taken in the past, and this provides a rough yardstick against which to measure performance.

(b) Orders may be given in many different forms, e.g. verbally, by letter, printed form, etc., and standardisation of requests is a necessary prerequisite.

4. Manhours per job

This method is similar to the previous one and has the same drawbacks. It may be used for frequently-occurring jobs of a similar size and character but the job unit is usually too variable to provide a basis for comparison.

5. Comparison with estimated times

$$\frac{\text{Total estimated manhours on jobs}}{\text{Total actual manhours worked on same jobs}}$$

(a) The value of the results obtained will depend upon the accuracy of the estimating although, if consistent, the ratio will at least indicate the movement in efficiency. However, estimating accuracy may vary from trade to trade with the result that the ratio will be affected by the mix of trades in any period.

(b) It is assumed that the estimate accurately represents the difficulties of the particular job whereas in many cases the estimated time is an average.

(c) A proportion of the work may not be capable of being pre-estimated and post-estimates prepared for this purpose may be influenced by the time actually taken, or deliberately devised to give a certain rate of bonus.

6. Standard hours

This is a more sophisticated form of the above method in which the work content is expressed in standard hours, i.e. the amount of work that can be performed in one hour by a worker of representative skill and experience motivated by a suitable incentive and with allowances for relaxation and other contingencies. Ideally, the times should be obtained from work measurement but, if this is impracticable, by estimating based on past records or by experience. The measure may be expressed in different ways, e.g.:

$$\text{Performance factor} = \frac{\text{Standard hours of work produced}}{\text{Actual hours expended on the work}}$$

$$\text{Cost per standard hour} = \frac{\text{Total cost of labour employed}}{\text{Standard hours of work produced}}$$

Bampton[5] suggests that the common unit should be the gross cost of one hour's actual work on site. The method involves finding the percentage of non-productive time using activity sampling techniques and then deducting this from the total hours worked to arrive at the number of productive hours. The total labour cost for the job including bonus and overheads is then divided by the resulting number of productive hours.

$$\text{Gross cost of 1 hour's actual work} = \frac{\text{Wages} + \text{overheads} + \text{bonus}}{\text{Actual hours worked} - \text{non-productive time}}$$

7. Accounting ratios
A rough indication of the variations in maintenance costs may be obtained by expressing the total annual maintenance costs as a percentage of replacement value or some primary function of the firm, e.g. production labour costs, total occupation costs, sales, etc. However, in all these cases there are other factors apart from maintenance efficiency which will affect the percentage. For example, the ratio between maintenance costs and replacement value (assuming a realistic figure is calculated annually) will be affected by the disparate rates of increase of maintenance and new construction costs.

Measures of planning efficiency
These are complementary to the productivity measures, which

they seek to explain in terms of the consequences of management decisions. They all presuppose a planned approach to the organisation of maintenance work and indeed are only meaningful in such a context. The measures used for this purpose include:

1. Degree of scheduling

$$\frac{\text{Total direct hours on schedule}}{\text{Total direct hours available}}$$

Clearly the aim of a planned maintenance system must be to predict and schedule as large a proportion of the workload as possible. The many small day-to-day routine tasks may be scheduled by the total times per period rather than individually. Variations in the ratio therefore indicate either a different proportion of small jobs from that anticipated or that larger jobs had not been foreseen.

If the proportion of scheduled work is too low, the individual jobs should be examined to ascertain whether or not they could reasonably have been predicted. The cause may have been failure to detect incipient faults at the time of carrying out inspections, too long an interval between inspections, or extraneous causes such as exceptional weather conditions or abnormal use. Conversely, a high proportion of scheduled work might indicate that preventive maintenance is being carried out at too high a level and that some work is being done prematurely and therefore at needless expense.

2. Lost time factor

$$\frac{\text{Actual hours productively employed}}{\text{Total time worked}}$$

This ratio will have meaning only if the causes of lost time are made explicit and, in particular, if a distinction is made between time lost as a result of inefficiency and that due to natural causes outside the control of both the operative and management. The avoidable causes (or at least those which should be examined to see if they were in fact avoidable) should be subdivided into:

(*a*) Waiting for instructions.

(*b*) Waiting for materials.
(*c*) Waiting to gain access to premises.
(*d*) Travelling time.

3. Supervision factor

$$\frac{\text{Total supervision costs}}{\text{Total direct labour costs}}$$

Alternatively this could be extended to cover the total management and supervisory costs. The introduction of a planned system, especially if coupled to an incentive scheme, will inevitably lead to higher management costs, but the increase should be geared to the consequential savings resulting from improved productivity.

4. Incentive coverage

$$\frac{\text{Total manhours on bonus work}}{\text{Total direct manhours available}}$$

The object is to give some indication of the extent of the coverage of the incentive scheme. Ideally, where there is an incentive scheme it should be applied throughout to avoid operative dissatisfaction. However, it is in the nature of maintenance that some of the work will be of uncertain scope and not capable of accurate pre-estimation for the purpose of fixing a target. In such cases, post-targeting may be resorted to, but this is unlikely to provide a stimulus to increase productivity. Also, it is a misuse of the word 'target' if the time is not determined until after the work has been executed, when it will tend to reflect the time actually taken rather than the time which should have been taken.

Service efficiency

Maintenance provides a service to the users of the building, and high productivity and planning efficiency are of little avail if the service is unsatisfactory. The following measures reflect the quality of the service provided.

1. Delay in executing work orders

The user of the building will tend to judge the efficiency of the maintenance organisation by the promptness with which his requests are dealt with. This may be indicated in various ways:

(a) By the average time taken to respond to authorised requests during the period under consideration.

(b) By calculating the delay ratio:

$$\frac{\text{Number of jobs one week overdue}}{\text{Number of jobs completed in same week}}$$

(c) By the backlog of work orders accumulated up to the end of the period in question. A breakdown of the backlog by trades will also indicate the adequacy of the labour force available and assist in manpower planning. A small backlog may indicate that necessary maintenance is not being reported or that maintenance manpower is at too high a level.

In interpreting these ratios the trend is more significant than the weekly figures.

2. Complaints from users

Such complaints may be made in many different forms and may be frivolous or refer to matters of basic design which are outside the scope of maintenance. It would be difficult to combine them into a single measure of dissatisfaction but they may point to specific defects in the system which require examination.

3. Consequential costs of breakdown

This applies to commercial and industrial buildings in which a breakdown may interfere with production processes or necessitate the closing down of part of the building while emergency repairs are carried out. The costs flowing from a breakdown are just as important as the cost of the remedial work when determining the justifiable amount to spend on preventive maintenance.

External agencies

These are the agencies which affect the cost of maintenance and which are outside the control of the maintenance organisation.

1. Cost of employing labour

$$\frac{\text{Total cost of maintenance labour}}{\text{Total direct manhours applied}}$$

The 'all-in' cost per hour will reflect changes in the basic wage rate and the various statutory and other levies which the employer is required to pay. The cost of supervision should not be included, as this is within the control of the maintenance organisation and is better considered separately.

2. Cost of materials

The movement in materials prices could be represented by a single index which is appropriate to the particular mix of materials normally used by the organisation, i.e. a compound index derived from the price indices of the individual materials weighted according to their proportionate use. However, the maintenance organisation can exercise some control over this factor by buying in a cheaper market or in larger quantities, or by using cheaper substitute materials.

3. Time lost due to unavoidable causes

Here the main factor is bad weather, although on examination of the circumstances it may be found that some of the time lost could have been avoided by providing alternative indoor work. This measure would be a subdivision of the lost time factor mentioned earlier.

The measures considered so far have been concerned with efficiency and with minimising the use of resources. It is equally important to consider the effectiveness of the work in achieving the basic objectives of the organisation; also whether or not the work was necessary or appropriate, and the quality of its execution.

Cost indices

Indices are used for the purposes of updating historic cost

data, for estimating and adjusting the costs of work carried out at different times to a common level for comparison.

Index numbers provide in a single term an indication of the variation against time of a group of related values. The simplest case is where it is required to compare the changes in price level of a single item over a period of time. A specific year is chosen as the base year and the price in that year equated to 100 per cent. Prices in later years are then expressed as percentages of the base year price. When stating the index number the percentage sign is omitted, as shown in Table 5.1.

Table 5.1 Example of cost index: bricks

	Price per 100 £	Index number
1975	15·00	100 (base year)
1976	18·75	125
1977	22·00	147
1978	25·00	167
1979	27·00	180

In many cases it is necessary to reflect the general movement of complex variables which consist of groups of items which have different variations. An example of this would be an index to reflect the changes over a period of years of the cost of carrying out maintenance work. One way in which this can be done is to analyse the work over a period of time and identify the major components of cost. This involves determining the proportions of the total cost attributable to labour and each of the principal materials. Then by weighting the separate indices for labour and materials according to their relative importance a combined index can be obtained.

Example

It is required to determine the average increase in the basic cost of maintenance work from 1975 to 1979.

Calculation of Combined Index for Materials

Material*	Index for 1979	% rise over base year	Base year 1975 = 100 Weight (% of total annual cost of materials)	Product
Timber	169	69	15	1035
Slates and tiles	182	82	5	410
Cement	197	97	10	970
Plaster	164	64	10	640
Glass	176	76	2	152
Paint	160	60	33	1980
Bricks	180	80	10	800
Copper pipe	194	94	3	282
Sanitary fittings	188	88	3	264
Ironmongery	185	85	2	170
Sand and gravel	214	114	2	228
			95†	⟩6931

Average % increase in price
of maintenance materials   =   73
from 1975 to 1979
Index for 1979 = 173

---

* Price indices for individual materials have been taken from housing and construction statistics published by the Government Statistical Service.

† Remaining 5% miscellaneous materials.

The total average increase of labour and materials can be found in a similar way.

	Index for 1979	% increase over base year	Weight (found by analysis of annual costs)	Product
Labour	172	72	65	4680
Materials (weighted)	173	73	35	2555
			100	⟩7235

Average % increase in price
of maintenance labour and   =   72
materials from 1975 to 1979

Index for 1979 = 172

A general criticism of this type of index is that it has to take into account a number of accumulative factors that are difficult to measure and combine. In particular, the productivity of labour which has an effect on the real cost of labour is difficult to assess. Also the pattern of costs is likely to change over a period of time and as this will affect the weighting structure of the index it will be necessary to start a new series when the change becomes pronounced. This might well be accompanied by the introduction of new materials which did not figure at all in the original list and by a decline in the importance of other materials which previously had a significant effect on total cost. The adoption of alternative materials in this way might be the result of price increases in traditional materials and would necessitate a change in the weightings of the individual indices to reflect the effect on the combined index of such innovations. It will be appreciated that an index is only an indication of the average movement of prices and cannot reflect the particular difficulties surrounding an individual job.

Sometimes it is necessary to change the base year, either in order to update the series or to compare two series which have been computed according to different base years. For example, assume that the old series has 1970 as the base year and it is desired to compute a new series based on 1975.

Bricks Year	1970	1971	1972	1973	1974	1975
Old index	100	115	132	148	180	220
New index	45	52	60	67	82	100

The procedure is to look up the index number relating to the new base year, i.e. 220 in the example, and divide this number into each index number in the old series and multiply by 100. This will give a new series of index numbers with the new year as base. However, changing the base of a weighted index in this way gives a series which is slightly different from that which would have been obtained had the series been computed completely afresh but for practical purposes the difference is rarely significant.

An alternative method of constructing an index is to base it on the recorded costs of carrying out a standard unit of work at different points in time. This can be applied where the same job has to be carried out at regular intervals under similar conditions and has a constant work content. It is essential that the same variable should be recorded from time to time in order to give a true comparison. For this reason the method is most suited to well-defined single-trade jobs such as painting, where a typical job would be selected and the quantities of the financially significant items measured and priced out at standard rates. At intervals the items would be repriced at the rates prevailing at the time and the total job cost so obtained compared with the cost derived from the standard rates. The costs at the various points in time would then be related to the standard rate cost which would be equated to 100. The index would, of course, only be applicable to the particular trade or job under consideration which would be taken as representative of the overall workload of the organisation. The index would reflect not only differences in labour and materials prices but also such factors as efficiency of maintenance planning, materials purchasing and labour productivity.

A simpler method is possible where work is let to a builder on the basis of a pre-priced schedule of rates, in that the percentage additions on the schedule rates quoted by the builder can form the basis of the index. It is, however, difficult to produce a reliable general index for maintenance due to the great variability in the conditions under which the work is executed and the wide range of prices charged by contractors for apparently similar work.

Quality control

In minimising cost, it is important to ensure that the quality of the work done does not suffer. This is a supervisory function and is performed by inspecting work during execution and on completion to check that the materials used and standard of workmanship are appropriate. It should be noted that whereas for contract work there is usually a clause in the contract requiring the contractor to make good defects appearing within a specified period, there is no such remedy for defective work done by directly employed labour. The important factors are as follows.

1. Creation of conditions which will favour good quality work such as:

(a) Selecting a reputable contractor of known reliability.
(b) Employing operatives with the requisite skills.
(c) Planning the work so that it is not affected by adverse weather conditions.
(d) Arranging, as far as possible, unimpeded access and suitable working space and conditions.

2. Clear instructions as to what has to be done and how. Where a specification is provided for contract work it should describe the various criteria which must be met if the work is to be judged acceptable. Care should be taken to ensure that the specification requirements are in fact practicable.
3. Inspections during the execution of the work timed according to:

(a) The complexity of the work − visits should be timed to coincide with the execution of difficult parts of the work.
(b) The need to amplify the written instructions − this would apply where the full scope of the work is ascertainable only after preliminary work has been carried out.
(c) The time-span of discretion related to the skill and reliability of the operatives concerned.

The object should be to inspect at critical stages in the progress of the work, to avoid a situation where work has to be re-done with the consequent waste of effort and material.

References

[1] Chartered Institute of Building (1983). 'Code of Estimating Practice.' 5th edition. Ascot: Chartered Institute of Building.
[2] Jarman, M. V. (1967) 'Selling maintenance to management − the use of proper costing.' London: Conference on Profitable Building Maintenance.
[3] Ray, H. G. (1969) 'Budgeting for Maintenance.' London: Conference on Building Maintenance.
[4] Staveley, H. S. (1967) 'The Planning of Maintenance.' London: Conference on Profitable Building Maintenance.
[5] Bampton, E. (1967) 'Direct labour or contract.' London: Conference on Profitable Building Maintenance.

CHAPTER SIX

Information system

Functions of system

Kochen [1] has described the functions of an information system as 'planning the behaviour of an organism, of alerting the organism to changes in its environment that signal the need for action, and of controlling action towards the implementation of a plan'. A management information system has been defined by the Institute of Cost and Works Accountants [2] as 'a system in which defined data are collected, processed and communicated to assist those responsible for the use of resources'.

Many writers have commented on the paucity of information relating to maintenance; the following extract from a conference paper is a typical example: 'It is regrettable that the scientific collection, assessing and analysing of data relating to problems of maintenance and repair is very rarely carried out and information when available is frequently so coloured by a particular situation or prejudice as to be of doubtful value.' However, the collection of information is not an end in itself and is only of value if the information is applied to control actions towards the achievement of specific aims dictated by policy considerations. It is important, therefore, that the basic aims should be clearly identified and appropriate procedures and techniques developed for the storage, retrieval and processing of relevant information. As stated by Robertson [3] information and information flows are justified by their relevance to decision-making and the action which has to be initiated. He

points out that the earlier the decision is made the more important and irrevocable are its financial consequences, and yet it is
at this stage that there is usually the least information available.

It is necessary to examine the needs of the users of the
information, the nature of the information they require, the
sources from which it may be obtained and the mode of presentation which will facilitate its use in the decision-making
process.

Information needs of users

In order to ensure a consistent approach it is desirable that
the basic information collected should be capable of being
processed to meet the particular requirements of decision-
makers at different stages in the building system. However, in
this context only the information needs of those decision-
makers within the owner/occupier and design systems will be
considered.

Owner/occupier system

Two sub-systems can be distinguished:

1. Maintenance expenditure control (see Fig. 6.1)
This is basically a function of upper management and is concerned with the total amount to be spent over a period of time
rather than with the detailed costs of individual items, unless
these happen to be large and non-recurring. For this purpose a
fairly coarse yardstick would be satisfactory against which to
measure the reasonableness of the expenditure proposed or
incurred. As stated by the Working Party on Local Authority
Housing,[4] 'We do not think that any properly managed organisation, whether public or private, would incur the management
and maintenance expenditure of even a fairly small housing
scheme without wanting to know how its expenditure compared with that of other organisations operating in the same
field'.

However, it would be unwise to place too much reliance on
costs incurred by other organisations in which circumstances

might be quite different. The person exercising the function of maintenance cost control is subject to various pressures and his decision to authorise a certain level of expenditure on maintenance will be affected by the strength of the demands from other quarters and, in part, by his own interest in and knowledge of

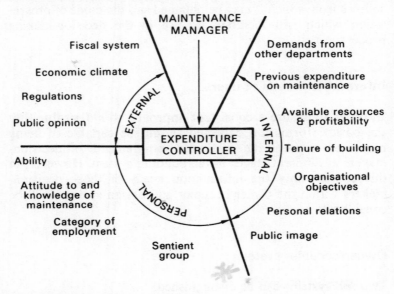

Fig. 6.1. Maintenance expenditure control

maintenance. It could be a highly individual decision and unlikely to form a reliable basis for other firms in which conditions and financial pressures are different.

2. Maintenance management

The functions in this area are mainly of a technical nature and concerned with the planning and control of construction resources to ensure that necessary repairs and renewals are carried out with maximum efficiency and economy. The major decisions relate to:

(a) Determining standards. For this it is necessary to have information on the overall objectives of the organisation and of statutory and other external requirements so that compatible standards can be fixed. The expression of these standards in qualitative and quantitative terms demands knowledge of the effects of varying degrees of disrepair on user activities and levels of visual acceptance.

(b) Planning inspections. Fixing the periodicity of inspections requires knowledge of the rates of deterioration of the building elements so that defects are revealed before they have reached a critical stage. The minimum period will be determined by the inspection costs which should clearly not exceed the cost consequences of failure.

(c) Identifying and specifying the work necessary. This is achieved by comparing the information received on the condition of the building from inspections and other sources with the standards laid down. It demands knowledge of the causes of defects and of the remedial measures which would be appropriate in the circumstances. Often the assessment is based on accumulated experience but there are obvious dangers in applying traditional remedies to new constructional forms and materials.

(d) Estimating the cost of the work. As far as possible the estimates should be based on historic cost data obtained from within the organisation for previous similar jobs, but in the absence of such data, costs from external sources and experienced judgment have to be used.

(e) Planning the work. This is mainly in respect of fixing appropriate start and finish times for the individual jobs and requires information on the effect of the timing of the work on user activities, its urgency, the availability of resources and the labour time required for each operation.

(f) Organising the execution of the work. The major decision in this area is whether to employ labour directly for the purpose or to engage an outside contractor: for this, information will be required on the relative merits of these alternatives from the point of view of both cost and convenience. It will include:

(i) For contract work, information on contractors' price levels, seasonal variations, and the cost and quality implications of alternative tendering procedures and contractual arrangements.

(ii) For directly employed labour, detailed information on wage rates and labour on-costs, output 'norms', job methods, union agreements, prices of materials and plant charges.

(*g*) Controlling cost, performance and quality. This involves formal systems for the feedback of information on progress so that actual costs and performance can be compared with those predicted and remedial action taken if necessary.

The relationships between the above functions are illustrated in Fig. 6.2.

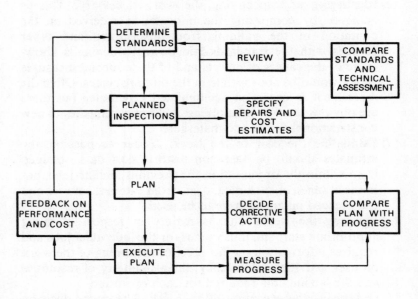

Fig. 6.2. The relationships between maintenance management functions

Design system

The design system embraces all those who contribute to producing a design solution which meets the client's requirements in spatial, environmental and cost terms. In addition to the usually accepted members of the design team, i.e. architect, quantity surveyor, structural engineer and mechanical and electrical engineers, one should include the maintenance manager for advice on the maintainability of the design. Their information needs in relation to maintenance include:

1. Knowledge of the client's maintenance policy. This relates to general policy in that the details depend on the design finally produced. The probability that repairs will be carried out promptly will influence decisions concerning the required durability of the various elements and components. Where the client owns other buildings an examination of these will give a general indication of his attitude towards maintenance. Also the employment of a direct labour force and the operation of a planned maintenance system would suggest that periodical servicing is likely to be attended to. An important factor which is rarely mentioned in the brief is the planned life of the building and yet this is essential information if the designer is to design for minimum maintenance and running costs commensurate with the initial capital cost.
2. Information on the behaviour in use and maintenance characteristics of alternative materials and design solutions. Surprisingly there is very little hard information in this field and of that which does exist only a small proportion is presented in a form which is easily and conveniently usable.
3. Knowledge of the way in which maintenance and cleaning will be carried out. This information is required so that suitable access can be provided and proper safeguards incorporated to protect the operatives when carrying out the work.
4. Information on maintenance costs for the purpose of calculating the total costs – i.e. initial capital cost plus subsequent maintenance and running costs – of alternative designs. It is doubtful whether actual job costs would be of very much value in this context, since they are affected by many factors other than the intrinsic properties of the design element, e.g. working conditions, tender climate, management skills, etc. It is more important to have some guide as to the frequency with which repairs are necessary and the probable lives under different conditions of the components under consideration.

Uses of information

It is clear from the foregoing examination of needs that information is used for three main purposes – prediction, comparison and knowledge (or instruction).

Prediction

This involves the collection of data over a period of time in order to establish trends or the conjunction of information to establish a cause-and-effect relationship, e.g. the bringing together of information on the properties of materials and the environmental conditions to which they will be exposed to forecast their probable behaviour in use.

The nature of the information and the most suitable form of presentation will depend upon:

1. What has to be forecast, the reasons for the forecast and the amount of detail necessary.
2. The time scale of the predictions and the degree of accuracy necessary.
3. The sources of the information, its reliability and the frequency with which new information becomes available and whether the new information supersedes the old or enlarges it.
4. The importance of external factors and the ways in which these can be anticipated and their effect measured.
5. The impact of the predictions on other parts of the system and the overall effect of prediction errors.

Comparison

This may take the form of comparisons of internal data for the purpose of measuring performance or of comparing internal data with that obtained from external sources as a check on efficiency. It is important to ensure that the sets of data being compared have been collected and classified on the same basis and that the background conditions are similar.

Knowledge/instruction

In many cases the information is required for the purposes of explaining what has to be done and how it is to be done. The amount of detail and form of presentation will depend upon:

1. The purpose for which the information is required. In some cases the information may be solely for internal use within

the firm while in others it may be contained in documents which form part of a binding contract, e.g. drawings, specifications and bills of quantities.
2. The knowledge already possessed by the recipient. Documents rarely contain all the information required by the recipient for an activity and assume a sufficient background knowledge to fill any gaps.
3. The method of communication. Although the primary means adopted are graphical and written, a good deal of information is conveyed verbally and it is anticipated that greater use will come to be made of punched or magnetic tape.
4. Economy. The production and transmission of information should be efficient in the use of the resources of the originator and of the recipient.
5. Multi-use of information. In addition to the initial instructional use of contract documents, these also provide a source of information for predictive and comparative purposes.

Sources of information

There are two points to consider: firstly, whether the amount of information is adequate and, secondly, whether the information that is available is being properly used. Although attention is frequently drawn to the lack of maintenance information there is quite clearly a vast amount in existence which, for a variety of reasons, is inadequately used. Perhaps the main reason is the difficulty in finding relevant information at the time it is required. Faced with the plethora of unco-ordinated information, the decision-maker tends to use only the most accessible material, combined with intuition. The information may be obtained from internal records or from outside sources.

Internal sources of information

1. Statements from upper management. It is essential that the maintenance manager should be apprised of the overall policy in so far as this influences the standard of maintenance. This information is required in order to assess the extent to which different standards affect user activities and

either hinder or advance the achievement of the organisation's overall objectives. It may be that where maintenance expenditure forms only a small proportion of total expenditure upper management does not think it worthwhile to analyse the problem and is content to leave decisions solely to the technical judgment of the maintenance manager. However, this inevitably leads to an *ad hoc* approach to maintenance and removes the incentive to introduce a planned system designed to secure the most economic use of resources.

2. Building inventory/project information manual. This document lists the facilities to be maintained, their location and condition, and is an essential pre-requisite to planning maintenance. For new buildings a maintenance manual may be provided by the designer, but will require continuous revision in the light of direct experience of the building, and to incorporate any alterations or additions.

The Maintenance Manual and Job Diary published by the Building Centre[5] is divided into three parts, as follows:

Sources of information	Description of building
	Contract consultants
	Contract information
	Authorities
	Subcontractors and suppliers
	Emergency contracts
	Schedule of floor areas and loadings
	List of maintenance contracts
General maintenance	Regular cleaning
	General maintenance guide charts
	General maintenance instructions
	General maintenance log sheets
	Fittings for replacement
Services maintenance	Services maintenance guide chart
	Fittings for replacement
	Log sheets
	Drawing list

The manual is prepared by the designer largely on the basis of information received from manufacturers and suppliers. If properly prepared, it can be a most useful document but in many cases the maintenance and inspection periods are not related to any defined maintenance policy and are therefore unlikely to be the most economic. Also, in practice, one finds that only a proportion of the items are listed in the manual and that it falls far short of being a comprehensive planned maintenance system.

3. Inspection reports recording the actual physical condition of the building. The quality of the information will depend upon the skill of the inspector and the degree of forethought given to the setting of test criteria and procedures. The preparation of standard forms to serve as check lists is an important means of guarding against omissions. A typical form from a government publication[6] is reproduced in Appendix A.

4. User reports relating to inadequacies and faults which detract from the user of the building. Generally information from this source will be incomplete and will require previewing by a member of the maintenance organisation to establish the precise nature and scope of the work involved. Again, it is desirable to standardise the form in which such requests are made and the reporting procedures.

5. Documents used for the initiation and control of work. The work order is the major source of information on resources used by directly employed labour. Against the standardised description of the job the completed work order should indicate the number of labour hours per craft and the materials used. Suitably coded, the information can be processed to give costs per building, per department, per element, per craft, etc., and if necessary may be subdivided into labour and materials.

6. Contract documents such as priced bills of quantities, schedules of rates and accounts submitted by contractors on completion of jobs. The amount of information which it is possible to elicit will depend upon the extent to which the work is broken down for the purpose of tender pricing. For new construction, the priced bill of quantities forms a valuable source of information for cost planning, but this is a very detailed and standardised breakdown into small items of labour and material which are priced individually. For maintenance work there is no similar standard approach to the preparation of tender and contract documents, and in

many cases the contractor merely quotes an all-in lump sum or carries out the work on a cost reimbursement basis. In such cases only the total contract sum is recorded and this is not capable of subdivision or adjustment for the purpose of estimating the cost of future jobs.

External sources of information

These are extremely varied and the following account indicates only the more important. Apart from textbooks, the sources include:

1. Government departments

(a) The Department of the Environment in conjunction with the Government Statistical Service publishes statistical data of, among other things, the composition of the stock of buildings, the input of the construction industry in terms of type and size of firms, operative trades and earnings and the output by type of client and building.

(b) The Property Services Agency was set up within the Department of the Environment in 1972 to advise government departments on property management, construction and maintenance. The Agency has a library service from which publications on various aspects of construction and maintenance can be obtained.

(c) Other government departments issue design guides and recommendations on maintenance procedures for the types of building for which they are responsible. The publications of the Department of Health and Social Security are particularly helpful in this respect.

2. Other sources

(a) The Building Research Establishment has for many years carried out research into and published papers on the technical, economic and organisational aspects of construction and maintenance. Of these publications the Digests (First and Second Series) give useful and concise information on the causes of defects and preventive and remedial measures.

(b) The Chartered Institute of Public Finance and Accountancy provides a professional accountancy body for the public sector. The Institute publishes annual statistics of the maintenance costs of local authority housing and also reports on accounting procedures that may be used for this type of work.

(c) The British Standards Institution publishes specifications and codes of practice for materials and workmanship respectively. Unfortunately, few British Standards involve a requirement that there should be experience of satisfactory performance in use and/or the ability to withstand laboratory tests. Also, they lag behind current practice and it takes a considerable time before new materials or methods are incorporated in a BS specification or code of practice.

(d) The British Board of Agrément was set up in 1966 to provide technical assessments of new building products. The certificates of approval include an assessment of the probable performance of the product in its intended use and in appropriate cases the maintenance requirements and probable life. They are therefore more informative than BS specifications in regard to the maintenance characteristics of materials but at present cover only a limited range of products.

(e) Manufacturers' literature is produced in very large quantities and varies greatly as to the quality of the information given. Although untrue statements are discouraged by the Trade Descriptions Act, 1968, the assessment of claims made for the performance of products often requires skills and equipment not usually available to the designer.

(f) Trade associations have been formed to promote the proper usage of materials by their member firms; most of these associations publish information on maintenance methods and are willing to advise designers and maintenance managers on particular problems.

(g) LAMSAC (Local Authorities Management Services and Computer Committee) is a specialist organisation set up in 1968 to provide a central advisory and consultancy function on management services in local government. In its advisory role it provides general advice to its members and training in the techniques of management. It evaluates and develops new systems in management and computing and provides a reference source to published information and data.

(*h*) SLASH (Scottish Local Authorities Special Housing Group) has published a ten-section manual of maintenance practice to assist those responsible for the supervision and management of housing maintenance. The guide divides housing elements into ten major sections and considers each section with regard to planning, specification, cost levels and feedback.

(*i*) The Chartered Institute of Building is actively engaged in the furtherance of maintenance studies and for some years has had a special committee to examine the problems in this field. In addition to an extremely useful booklet entitled Maintenance Management – A Guide to Good Practice, their maintenance information service has published a number of papers on specific maintenance topics and an up-to-date review of the available literature on the management and organisation of building maintenance. The Institute also offers a diploma examination in Building Maintenance Management.

(*j*) The Building Maintenance Cost Information Service was formed in 1971 to provide a focal point for the collection and dissemination of information on the technical, organisational and financial aspects of maintenance. It offers a means whereby subscribers can exchange their knowledge and experience. One of its primary objects is to encourage better communications between upper management and maintenance management and between property users and the design team. The information is presented under the following heads:

General and background information. This section includes news articles, trends in prices of labour and materials, wage rates and labour on-costs, development of materials and equipment, new legislation and statistical series appertaining to maintenance.

Publications digest. This section gives brief summaries and sources of articles relevant to property occupancy.

Case studies. The reported cases cover aspects of maintenance management and budgeting and control systems.

Occupancy cost analyses. The cost analyses cover the whole range of occupancy costs and are prepared from data supplied by subscribers. They are presented in a standard form as follows:

0 Adaptations
1 Decorations
   1.1 External decoration
   1.2 Internal decoration
2 Fabric
   2.1 External walls
   2.2 Roofs
   2.3 Other structural items
   2.4 Fittings and fixtures
   2.5 Internal finishes
3 Services
   3.1 Plumbing and drainage
   3.2 Heating and ventilating
   3.3 Lifts and escalators
   3.4 Electrical services
   3.5 Other M & E services
4 Utilities
   4.1 Gas
   4.2 Electricity
   4.3 Fuel oil
   4.4 Solid fuel
   4.5 Water rates
   4.6 Effluents and drainage charges
5 Cleaning
   5.1 Windows
   5.2 External surfaces
   5.3 Internal
6 Administrative costs
   6.1 Service attendants
   6.2 Laundry
   6.3 Porterage
   6.4 Security staff
   6.5 Rubbish disposal
   6.6 Property management
7 Overheads
   7.1 Property insurance
   7.2 Rates

Design/performance data. The object is to describe the state of elements and components, to examine the causes of failure and to suggest the design correction that might have avoided the failure.

Research and development papers. These are papers on various aspects of property occupancy either by the service or by subscribers.

Reference indexes. The intention is to provide a cumulative reference index of key words to facilitate the retrieval of data.

More recently the BMCIS has introduced the following additional sections:

Legislation for property managers. This is designed to help maintenance managers to keep track of the ever-changing pattern of legislation by giving a summary of the statutory provisions which relate to the management of premises.

Energy cost analyses. These give details of the energy consumption of different types of buildings for the purpose of inter-firm comparisons.

Price book. This is devoted solely to the prices of maintenance work and should prove extremely useful to all who are involved in planning, budgeting, estimating and checking builders' accounts for this type of work.

The information sheets are sent to subscribers at regular intervals and eventually build up to form a comprehensive library of facts and figures on property occupancy.

Reliability and interpretation of external information

Apart from the fact that many firms do not keep their records in a suitable form or in sufficient detail for the purpose of maintenance management the information is likely to be uncertain for the following reasons:

1. There is no universally accepted definition of maintenance and the boundary between maintenance and improvement is often difficult to draw with any precision.
2. There is a lack of a universally accepted framework covering all divisions of maintenance with the result that maintenance costs from different sources are aggregated under different

heads and cannot be re-allocated without reference back to the source documents.

3. The heterogeneous nature of the work and the smallness of the individual jobs inevitably results in recording errors. Indeed, it is estimated that up to one-third of the annual output on maintenance is not recorded at all and of that which is recorded, the picture is distorted by such factors as different accounting systems or taxation considerations.

4. Most maintenance is carried out by small firms and as they are the least able to make use of statistical information, they have the least reason to take care in its initial collection.

However, the main problem lies in interpreting the data. The principal uncertainties are:

1. The data show only the amount spent and, as a rule, do not indicate:

   (*a*) The standard achieved.
   (*b*) Whether the work done was all that was technically necessary or whether part had been deferred for policy reasons.
   (*c*) The quality of the work done.
   (*d*) The efficiency with which the work was done.

   Clearly, low costs do not necessarily reflect efficiency; they could equally be the outcome of neglect. Indeed, abnormally low costs are more likely to stem from a policy decision to cut maintenance to the minimum rather than indicating that the building is in first-class condition.

2. The reasons for expenditure are not usually given and may be completely obscured by the aggregation of costs under inappropriate headings. For both the maintenance manager and the designer it is important to know whether the work was rendered necessary by faulty design or abnormal use or was attributable to normal wear and tear. Without this knowledge it is not possible to isolate exceptional costs or to understand why they were exceptional.

3. The costs of deferring work are not ascertainable from statistics of historic costs except in so far as high current costs may be traced to past neglect. In the context of planned maintenance it is essential to consider the cost pattern over a period of years and in particular the effect of alternative periods for cyclic work.

Information categories

1. Classification principles

Classification has been described [7] as the actual or ideal arrangement of those objects which are like and the separation of those which are unlike. The categories should be meaningful to the user and sufficient in number:

(a) to allow relationships to be built up between relevant groups of information, e.g. elements/standards/costs; maintenance task/resources/cost.
(b) to identify resources with regard to performance, availability and price.
(c) to facilitate the preparation of relevant documents, e.g. inspection schedules/work programmes/work instructions.

There are two basic types of information to consider:

General. Information which is freely available in published form and is useful for a variety of purposes.
Specific. Information which has a restricted application and which may be derived entirely from internal sources or consist of a unique collection of general information. It may relate to:

(a) The organisation owning or occupying the building, e.g. overall policy, management structure, financial situation, etc.
(b) The physical characteristics of the building(s), e.g. constructional form, floor area, etc.
(c) The individual items of maintenance work, e.g. material and labour resources, costs, etc.

The primary classification need for general information is to permit ready retrieval of the document, whereas for specific information the ordering should allow manipulation and processing for a defined purpose. However, it is possible for the same information to exist at both levels, e.g. a published standard schedule of work descriptions would be of general applicability whereas a unique permutation of selected work descriptions would be specific to a particular job. Ideally, therefore, both types of information should be contained within the

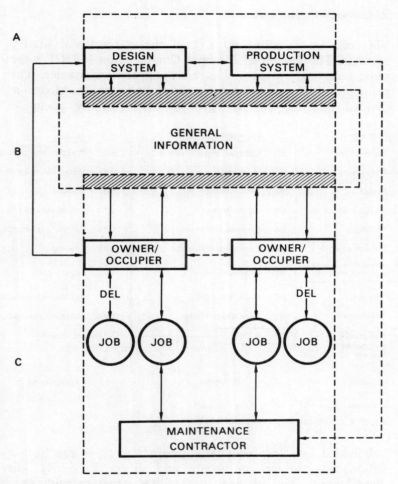

Fig. 6.3. Information flow
  A specific information relating to new construction system
  B general body of information
  C specific information relating to maintenance system
  Shaded area indicates overlap of general and specific information

same frame of reference. In view of the widespread use of existing classification systems for general information, it is probably more realistic to think in terms of a complementary but compatible set of categories for maintenance. Figure 6.3 indicates the overall pattern of information flow.

2. General documentation

This can be classified by any of the conventional library systems. The Universal Decimal Classification (UDC) is the most commonly used and is hierarchical in character. The principle employed is that the total body of knowledge is divided into ten primary groups, each primary group is

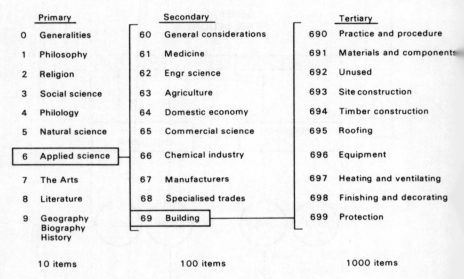

Primary		Secondary		Tertiary	
0	Generalities	60	General considerations	690	Practice and procedure
1	Philosophy	61	Medicine	691	Materials and components
2	Religion	62	Engr science	692	Unused
3	Social science	63	Agriculture	693	Site construction
4	Philology	64	Domestic economy	694	Timber construction
5	Natural science	65	Commercial science	695	Roofing
6	Applied science	66	Chemical industry	696	Equipment
7	The Arts	67	Manufacturers	697	Heating and ventilating
8	Literature	68	Specialised trades	698	Finishing and decorating
9	Geography Biography History	69	Building	699	Protection

10 items   100 items   1000 items

9 classes gives 10^9 = 1000 million items

Fig. 6.4. Universal Decimal Classification

subdivided into ten secondary groups and these are further subdivided into tertiary groups and so on (Fig. 6.4). The classification thus assumes a tree-like structure with each branch forking into a further series of branches as far as necessary.

The main disadvantage of this system is that the division into classes is fixed and does not permit regrouping under different heads. Thus there is a tendency for related material to be scattered throughout the classes and this creates difficulties when attempting to sort and compare data. Also the system must be conceived as a whole initially and it is difficult to anticipate how much space will be required for future developments.

Faceted classification

A faceted type of classification is usually more convenient for processing purposes. This creates a series of general concepts which, unlike the facets of a hierarchical classification, have equal weight. The advantages of the faceted system are that all possible combinations of characteristics can be catered for and that new concepts and characteristics can be added at any time.

Facet 1	*Facet 2*	*Facet 3*
Product	Colour	Finish
window	black	natural
door	white	primed
wall panel	red	decorative laminate
etc.	etc.	etc.

The concepts 1, 2, 3 ... are mutually exclusive and may be as numerous as necessary. Similarly the subdivision of each concept may be continued as far as necessary.

The SfB (Samarbetskommiten för Bygnadsfragor) is a specialised faceted classification which is widely used for filing technical literature relating to construction industry products. The system originated in Sweden, but is now under international control through the CIB (International Council for Building Documentation) which grants licences for publication through national agencies, e.g. the RIBA in the UK. The system was revised in 1968 and renamed CI/SfB. It includes four main tables:

Table 0	Built environment – a number, usually 2 or 3 digits.
Table 1	Elements – a bracketed number, usually 2 or 3 digits.
Table 2/3	Construction form and materials – a capital letter followed by a lower case letter and usually a number.
Table 4	Activities and requirements – a bracketed capital letter sometimes followed by a number and a lower case letter.

The references are entered in a box, as shown in Fig. 6.5. The detailed tables are given in the Construction Indexing Manual published by the RIBA, which also contains information on the method of choosing the correct symbols.

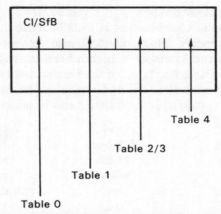

Fig. 6.5. Cl/Sf B classification: arrangement of reference tables

The tables relate mainly to new construction and the space allocated to maintenance is minimal. However, a good deal of the basic technical information is common to both aspects of construction activity and much of the literature from manufacturers and other sources is pre-indexed in accordance with the SfB classification. There would appear, therefore, to be some practical advantage in adopting SfB at the general level and extending it by the addition of specific maintenance facets.

It is clear that a fair amount of the information is capable of being presented in a standardised form and that transference of this information from the specific to the general level would facilitate communications not only within the maintenance system but also between the maintenance and the new construction systems.

Probably the most useful development would be the publication of standard work descriptions with related standard rates and labour outputs. This would provide a ready means of establishing trends in maintenance productivity and prices and give a more consistent approach to the preparation of contract documents and estimating.

3. Specific information

This is distinguished from general information not so much by its form as by its particularity. The main purposes for which it is used are the identification, timing, costing and execution of maintenance jobs. These activities both demand and generate information, some of which may be transferred eventually to the general system. The main groupings of this information are shown in Table 6.1.

Table 6.1.

	SPECIFIC			
GENERAL	Organisation	Building(s)	Facility	Operation
Technical	Objectives	Physical parameters	Performance standards	Description and method
Economic	Maintenance policy	Exposure conditions	Inspection criteria	Resource requirements
Management	Financial constraints	User activities	Inspection cycles	Timing
Legislation				Cost

Classification of maintenance costs

It is particularly desirable that the classification system should permit the grouping of maintenance costs in alternative ways to indicate:

Where   Building, department, use area, element or other appropriate locational reference.

What   The nature of the work done expressed in standard terms.

Why   The reason for carrying out the work and the cause.

When   The frequency at which the work becomes necessary.

How   The method of execution and, in particular, whether by directly employed labour or outside contractor.

However, although there are possible roles for a great many classes of maintenance cost, the number should be restricted to those which are essential in order to minimise recording costs. It should be borne in mind that many maintenance jobs cost less than £30 and that an over-elaborate system of classifying and coding may result in administrative costs which are excessive in relation to the value of the work actually done. The following classes of cost will be considered:

Location
Building element
Functional system
Purpose
Cause
Magnitude
Frequency
Source of funds
Job
Initiation
Execution

Location

Where the maintenance organisation is responsible for the upkeep of several buildings the system should allow allocation of costs to each building. Further subdivision into departments or use areas is also desirable. The usual accountancy practice is to distribute the total cost to departments on a floor-area basis, but this is not so meaningful as direct allocation. The physical parameters for each building should be recorded, i.e. age, form of construction, materials specification, number of storeys, floor area, use, etc., to form a basis for comparison with the costs of maintaining similar buildings.

The most meaningful locational reference for internal work is the 'room', which in this context would include all self-contained use areas, e.g. corridors and stairways. It has been suggested that greater use should be made of room schedules during the initial design as a basis for collecting information on sizes, finishes, windows, doors, fittings, services, etc.[8] These schedules could be developed progressively throughout the design stage and eventually incorporated in the Maintenance Manual.

This would give rise to two broad categories – the building envelope and the individual spaces which it encloses (Fig. 6.6).

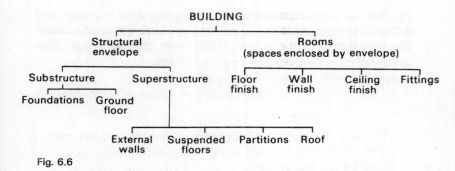

Fig. 6.6

Although different in kind, they reflect the parts of the building which are acted upon by the two primary causes of deterioration — for the building envelope (foundations, ground floor, external walls and roof) the effect of climatic factors, and for rooms (finishes to floors, ceilings and walls, fittings) the action of user activities. This would leave only the central services and external works to be dealt with separately.

Building elements

A building element has been described as a major component common to most buildings which usually fulfils the same function or functions irrespective of its construction or specification. The standard lists of elements produced for the purpose of cost planning new buildings have had a strong influence on the elements suggested for maintenance. It is pertinent, therefore, to consider how far the same form of analysis is suitable for both purposes.

The main uses of elemental analyses for new construction are:

1. To provide a reliable forecast of the cost of the building work.
2. To break down the cost in such a way as to enable decisions to be made regarding the development of the design.
3. To indicate probable user costs in total and with respect to individual elements of the building design.

It is only in respect of (3) that there is any direct identity of purpose. For (1) and (2) the elemental breakdown reflects the manner in which the designer conceives the building and develops the design of the parts. The primary difference is that whereas the designer is concerned with a hypothetical construct of a proposed building the maintenance manager is concerned with the physical result.

In the case of maintenance, the purposes of cost analysis are:

1. To reveal the pattern of expenditure in order to determine, for budgetary purposes, what is a reasonable amount to spend on the maintenance of each part of the building.
2. To identify high cost areas for further analysis and in particular to provide data to assist decisions as to whether it is more economic to renew a component or to continue to repair it.
3. To provide a check on the effectiveness of a planned maintenance policy by studying the incidence of costs over a period of time.
4. To assist in the prediction and control of future expenditure.
5. To compare the costs of like locations and elements both within and without the organisation.

For these purposes the elements chosen should meet the following requirements.

1. They should represent major cost areas so that time is not wasted recording trivia. However, it should be borne in mind that whereas it is always possible to produce broad analyses from highly detailed data it is not possible to produce detailed analyses from coarse-grained data and that although initially the fine details may be thought unnecessary there could be times in the future when they will be needed.
2. They should be capable of precise identification so that the costs can be accurately apportioned. Inevitably, there will be some small degree of overlap, e.g. where a floor enters a wall, and arbitrary rules are necessary to ensure consistency in the classification of work occurring at the junction of different elements. In laying down such rules, it should be borne in mind that the information is collected initially from sources close to the execution of the work and that therefore

the divisions should be natural ones which are easily recognisable on site.
3. They should be related to measurable parameters of the building so that costs can be adjusted for different conditions. Usually this does not go beyond expressing elemertal maintenance costs in terms of the amount per 100 m² of floor area. However, it is quite clear that many elemental maintenance costs are not directly related to floor area and that adjustments must be made for other factors if such cost analyses are to provide a reliable basis for inter-firm comparisons. The most important factors for which adjustments should be made are:

Quantity. It is apparent that a knowledge of the floor area alone does not indicate the actual areas of the elements to be maintained. Thus, a single-storey building will have ten times the roof area of a ten-storey building with the same floor area. The actual areas would provide a more reliable basis on which to make adjustments.

Quality. This relates to the condition of the building at the time and is the product of the initial specification and the age of the building. The age factor will tend to assume a constant value after about the first 15 years of the life of the building.

Environment. This would include climatic factors in relation to external elements and user activities in relation to the internal elements and finishes.

Policy. Although difficult to express in quantitative terms there is little doubt that the maintenance policy is often the dominant factor in determining the amount spent on maintenance.

4. They should be common to all buildings of the type under consideration. The more detailed the elemental analysis, the more difficult it is to satisfy this condition. However, commonality could be preserved at the higher levels of aggregation while allowing greater flexibility at the more detailed lower levels.
5. They should be related to maintenance management procedures and processes. Although alternative groupings may be necessary to suit the needs of particular procedures

it is desirable that the elemental breakdown should be compatible with these other groupings and form an integral part of the whole system.

6. They should allow logical subdivision and aggregation to suit the needs of different users, including feedback to the designer.

An examination of the published elemental analyses used for maintenance reveals that they fall into one of four basic groups:

(a) Those designed principally to facilitate inspections or field surveys for the purpose of combining an inventory of facilities and their condition. The elements are arranged in a convenient order for viewing and inevitably certain elements, e.g. the roof, have to be split into two parts because the interior and exterior are inspected at different points in time. It is usual for specifications of work to be done to be subdivided in a similar way. The following are two examples of this type of breakdown:

Bushell[9]	*Practice in Property Management*[6]
Internal	Internal
roof space	roof space
room interiors –	floors –
ceiling, walls, windows, etc.	ceiling, walls, windows, etc.
under-floor space	services
(services dealt with	
separately)	
External	External
roof exterior	roof coverings
roof plumbing	
walls, chimneys, etc.	walls, chimneys
doors and windows	doors and windows
	plumbing
external painting	(painting included with above)
drains	
fencing, walling	services and ancillaries
roadways and paved areas	
miscellaneous	

(*b*) Those which are an extension of normal accounting and cost control procedures, such as the following system described by Jarman.[10]

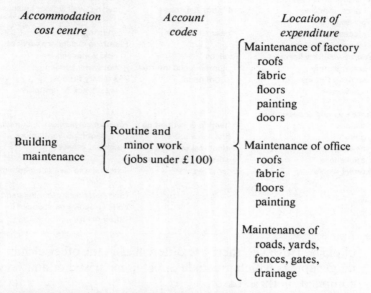

Accommodation cost centre	Account codes	Location of expenditure
Building maintenance	Routine and minor work (jobs under £100)	Maintenance of factory roofs fabric floors painting doors
		Maintenance of office roofs fabric floors painting
		Maintenance of roads, yards, fences, gates, drainage

(*c*) Those designed to permit comparisons of maintenance expenditure incurred by different organisations. In the three example analyses given in Table 6.2, decoration is recognised as being the major maintenance item and given pride

Table 6.2.

Housing[4]	Building Maintenance Cost Information Service[5]	Practice in Property Management[6]
	Decoration	
External decorating	external	External decoration
Internal decorating	internal	Internal decoration
Structure	Fabric	Main structure
walls, foundations,	walls	foundations & basements
		frame
chimney stacks		walls, chimneys, flues
roof	roof	roof
rainwater wastes		gutters & rainwater pipes
floors, stairs	other structural items	Internal construction
		floors
		stairs
		internal walls

Housing [4]	Building Maintenance Cost Information Service [5]	Practice in Property Management [6]
Structural finishings		Finishes & fittings
doors, windows	fittings & fixtures	ceiling, wall & floor finishes
glazing		
plastering & tiling	internal finishes	shelving, etc.
joinery & sundries		Plumbing & sanitary services
Water & sanitary services	Services	cold water service
water supply	plumbing and internal	hot water service
sanitary fittings	drainage	sanitary fittings
soil wastes		waste, soil & vent pipe
Other services & appliances		
electric	heating & ventilating	Mechanical services, including
gas	lifts, escalators	heating & ventilation
heating	electrical services	Electrical services
appliances	other M & E services	
External works	External works	External and civil engineering works
		(See secondary groupings and code structure reproduced in Appendix B)

of place. However, there are differences in the other elemental groupings which preclude direct comparison of analyses compiled on these bases.

(d) Those which parallel the elemental breakdowns used for cost-planning new buildings. These usually take the form of a detailed inventory coded so as to indicate the precise location of each individual item. These items are assigned inspection criteria and job descriptions for preventive maintenance and remedial work.

Functional systems

The term 'functional system' is used to describe the interactive result of a set of building elements in combining to provide some identifiable aspect of the user's environment. Building elements – walls, floors, roofs, etc. – simultaneously contribute to one or more functional systems. For example, the roof serves not only to exclude the weather but also by reducing heat losses forms part of the environmental thermal system. By relating costs to functional systems, the purpose of the work is more clearly demonstrated and it becomes more feasible to assess the effect on user activities of different levels of maintenance expenditure.

However, as this represents a new approach to the classification of maintenance costs, it would be better used as an addition to a system based on building elements to fulfil an explanatory function and to assist in relating costs to benefits.

A possible classification on these lines would be:

Safety (SAF). This would cover structural stability, fire precautions, security measures and hazards of all types. Usually these matters must be given first priority to comply with legal requirements, but some work, e.g. installation of a sprinkler system, may be carried out to obtain a reduction in insurance premiums.

Weather (WEA). This would include those parts of the external envelope which prevent the entry of wind and rain, e.g. roof coverings and wall claddings. The standard and priority will depend upon the extent to which the penetration of water will adversely affect other parts of the structure or interfere with the user of the building.

Environmental (ENV). This may further be subdivided into the visual, thermal and acoustic sub-systems and would include all work necessary to maintain the appropriate physiological and psychological conditions. It would include the service installations for heating, cooling, ventilation, lighting etc.; the contributions of the building fabric, internal decoration and cleaning.

Spatial (SPA). The internal subdivision of the building and the provision for movement between spaces both horizontally and vertically. This could be treated as a subdivision of Environment, but since spaces may have to be frequently rearranged to accommodate different activity patterns, it is better to consider such changes separately. It would also provide useful information on the value of flexible design methods and of provisions such as movable partitions.

Hygiene (HYG). This would cover washing and lavatory facilities and the associated waste and soil disposal systems.

Communication (COM). The audio and visual means of transmitting information including internal telephone, message tubes, closed circuit television, etc.

Where the installation is to protect the occupants or the contents of the building, e.g. fire and burglar alarms, it would be better classified under Safety.

Storage (STO). The spaces, shelving, cupboards, lockers, etc., for the storage of materials and equipment.

External (EXT). This would be subdivided into access (paths and drives), boundary protection (fencing and boundary walls), landscaping, drainage, supply mains for gas, water and electricity.

It will be appreciated that the conventional elements will fall into one or other of the above functional systems according to the purpose for which the work is carried out. Thus work to walls may be for the purpose of improving the insulation, in which case it would be classified under Environment, or may be to prevent the collapse of the wall, in which case it would be entered under Safety. The cost of the work can thus be related either to the benefits derived from an improved thermal environment (or reduction in fuel costs in achieving the same environmental conditions), or to a safer building.

Table 6.3 indicates the relationship between building elements, functional systems and maintenance operations.

Purpose

For the proper interpretation of cost data, it is essential to know the purpose for which the work was undertaken.

A useful sub-division would be into:

1. Repairs necessary to maintain the facilities to the required standard. Admittedly this does not resolve the problem of comparing costs incurred by different organisations, in that the base standards will vary from one building to another. An agreed method of grading standards is desirable so that costs can be adjusted to a common base.
2. Renewals. The decision to renew rather than to continue repairing may be influenced by economic or aesthetic considerations. It would be helpful to know which was the

Table 6.3.

Building element	Functional system	Operation
External walls	SAF	Repair/rebuild brickwork
	WEA	Patch/renew rendering
	WEA	Repair/renew cladding
	WEA	Remake joints between units
	SAF	Underpinning
Roof	WEA	Replace units/strip and renew slates/tiles
	WEA	Patch areas/strip and renew felt/asphalt
	WEA	Replace units/strip and renew sheeting
	WEA	Clean out gutters and rainwater pipes and remake joints
	SAF	Strengthen/renew structural timbers
	WEA	Repair/renew eaves fascia/soffit board/gutter boards
	ENV	Renew/provide insulation
Floors	SAF	Repair/renew floor finish
	SAF	Repair/renew structural timbers

determining factor, e.g. whether a component was renewed before the end of its functional life because it had become visually unacceptable.

3. Modifications. This would include alterations and improvements to allow a more effective use of the building and would be justified in terms of the predicted benefits flowing from the change.

4. Protection. This would apply principally to external painting and other thin coatings whose primary purpose is the protection of the base materials.

5. Decoration. This would include internal painting and other decorative treatments whose main purpose is to enhance the appearance of the internal surfaces of the building.

6. Cleaning. Although this also serves to maintain the internal appearance of the building, it is organised differently from normal maintenance work and is therefore better dealt with separately.

Cause

The cause of the failure is difficult to ascertain from the usual forms of analysis of maintenance data. And yet this is a key factor if past errors are to be avoided. Clearly the causes are many and varied and to record them all in detail would be exceedingly time-consuming. However, they fall into one or other of the following broad categories:

1. Normal wear and tear. This would be normal for the particular type of building, exposure conditions and user activities and does not imply that there is a certain level of expenditure which is applicable in all circumstances. However, it is suggested that where there is similarity of construction, exposure and user, it is logical to assume that the physical need for maintenance will be similar and that variations in amounts actually expended reflect differing policies and attitudes to maintenance.
2. Abnormal wear and tear. The abnormality may be due to:

   (a) Design fault. This may be in respect of constructional details or materials which are unsuited to the environment in which they are used.
   (b) Exceptional weather conditions for which it would not have been reasonable to make provision in the design.
   (c) Improper use. Where the possibility of rough usage could have been foreseen by the designer, failure to guard against it would amount to a design fault. But wilful damage or gross carelessness should be separately recorded so that where possible restitution can be sought or steps taken to avoid a recurrence.
   (d) Inappropriate maintenance. It would be unusual, perhaps, for maintenance management to enter anything under this heading, but there is little doubt that the wrong diagnosis of the cause of failure can not only fail to effect a cure, but can result in a worsening of the condition of the building.
   (e) Neglect. This is work made necessary through delay in carrying out maintenance at the proper time. Strictly speaking, the loss is the difference between the cost of the work done and the cost which would have been incurred

had the work been done at the right time. However, it is useful to draw attention to the higher costs resulting from delay and where the amount shown under this head is substantial, to reconsider the organisation's maintenance policy.

Magnitude

In most organisations one finds a distinction made between minor and major work. From a number of points of view – planning, authorisation of expenditure, source of funds – jobs are dealt with differently according to size. This may also be reflected in the manner of executing the work – direct labour undertaking the minor works, and the major works let out to contractors on a competitive or negotiated basis, with the consequential effects on documentation and supervisory and control functions. The division is usually achieved by fixing some arbitrary cost limit, e.g. all jobs up to and including £100 in value classed as minor works and those exceeding that amount as major works.

Frequency

This characteristic is usually linked to magnitude in that many minor jobs are of frequent occurrence, whereas major jobs, by their nature, tend to be infrequent. The value of this information lies in its use for predicting the incidence of future maintenance and for this purpose, it is important to distinguish between routine, cyclic and infrequent work. In particular, the periodicity and average cost of cyclic work should be established so that the effects of varying the period on total costs and on the user of the building can be ascertained.

Source of funds

The basic division is between revenue and capital according to whether or not the work increases the capital value of the asset. The distinction is significant with regard to tax, in that revenue expenditures may be set against taxable profits.

Job

Where the work is carried out by directly employed labour, the completed work orders showing time spent and materials used together with a standardised description of the job can be filed in numerical order for future reference. Although jobs rarely repeat themselves in precisely the same form, the completed work orders provide a useful aid to estimating and programming. Also, for control purposes, a schedule showing actual and estimated labour hours against each job number will indicate progress and efficiency.

Initiation

It is necessary for control purposes to have a measure of planning efficiency and this may be achieved by a breakdown into scheduled work, contingency work – i.e. that requested by the user – and emergency work. Generally, the higher the proportion of the scheduled work the more efficient the planning, although it should be noted that over-scheduling may be wasteful if it results in unnecessary work. The most advantageous proportions can be found by a careful analysis of the workload and will vary according to the predictability of the repairs and the nature of the occupancy.

Execution

Maintenance expenditure should be classified according to whether the work has been carried out by direct labour or by an outside contractor. Where direct labour has been used it is helpful for the purpose of manpower planning to subdivide into operatives' skills. Contract work may be subdivided according to the type of contract entered into.

Coding

Having decided upon the most useful categories of information to record, the next step is to devise a suitable coding system. In this context a code may be defined as a set of symbols used to facilitate the storage, retrieval and processing of information and data. The codes may be numerical, alphabetical or mnemonics of natural language words, e.g. BKWK for brickwork. Numerical codes are usually more convenient for machine and computer processing although an alpha numerical code, i.e. a combination of letters and numbers, is easier for visual recognition. The alphabetical code also has some advantage where the number of classes to be distinguished is greater than 10 but less than 26, in that only one letter is required instead of two digits. It should be borne in mind that the accuracy of the system will depend upon the correct allocation of costs on the job. The main source of error is in the recording of labour costs and this should be done daily by the foreman or supervisor. Clearly, the simpler the code structure the better from the point of view of reducing recording errors.

The processing of data may be carried out manually or by machine. Manual systems are time-consuming, costly and subject to human error. They are, however, flexible and once the procedures have been established they can be transferred to a machine. The final sophistication is to computerise the means of storing, sorting, calculating and presenting the information

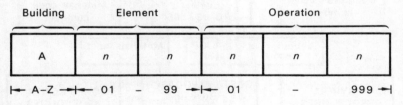

Fig. 6.7

in different formats for management control. Whatever system is used, the main source document is the completed work order showing the appropriate code symbols with details of labour hours and materials. For example, assuming that a maintenance organisation is responsible for the upkeep of a number of buildings and that it is required to classify costs according to location, element and operation, an appropriate code field would be that shown in Fig. 6.7.

Buildings. The buildings would be listed and each one assigned a letter, for example:

Office	A
Workshop 1	B
Workshop 2	C
Stores	D
etc.	

Element. A standard list of elements would be produced and each element and sub-element given a code number. For example, the roof might be 22.

Operation. Similarly a standard list of maintenance operations would be prepared and the operations numbered sequentially. For example, cleaning out gutters might be 156.

Thus the code B 22 156 would indicate cleaning out the gutters to the roof of Workshop 1. If all the work orders are coded in a similar way according to location, element and operation it is possible to obtain the total cost of maintaining each building, or the cost of maintaining individual elements or of carrying out a particular operation. The procedures would be as shown in Fig. 6.8.

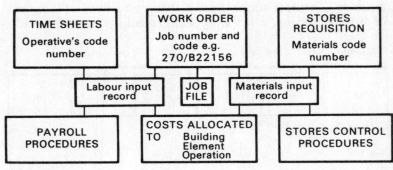

Fig. 6.8

Computer Applications

Hardware

Few technologies have developed at the speed of computer design or are as likely to have such a fundamental effect on our everyday lives. The first electronic computers were huge contraptions requiring special air-conditioned rooms and were extremely costly both to build and to maintain. Then, in the sixties, the invention of cheap logic chips made it possible to build computers the size of one or two filing cabinets for use in the small office. In accordance with the jargon of the time these were called minicomputers. Now with the development of high-density processor chips it has been possible to reduce the size still further to produce the desktop microcomputer and the even smaller portable models.

The heart of any computer is the Central Processing Unit (CPU) which is contained on a chip which has about a hundred thousand transistors arranged into a number of functional blocks. These blocks are an Arithmetic Logic Unit (ALU), some Random Access Memory (RAM) in which the ALU keeps the data it is working upon, some Read Only Memory (ROM) where it keeps basic operating instructions, a clock to control the speed of working and an Input–Output control and decode. The memory blocks are used to accumulate data to work on, to hold the programs that tell them what to do and to hold the results until they are passed on to the device which will use them. There is, however, only a limited amount of fast or main memory* built into the machine (the greater the amount of main memory the more expensive the computer) and therefore a second level of memory is kept on magnetic tape or disc. This second-level memory is cheaper than main memory but gives slower access. A typical computer configuration consisting of central processor and input, backing storage and output devices is shown in Fig. 6.9.

Operation of computer

Normal data processing involves the following activities:

Program input. A program prepared for the particular task is fed into the computer memory. This is held in the internal working store throughout the processing.

---

* This is sometimes referred to as core memory from the earlier method of using ferric rings for the internal store.

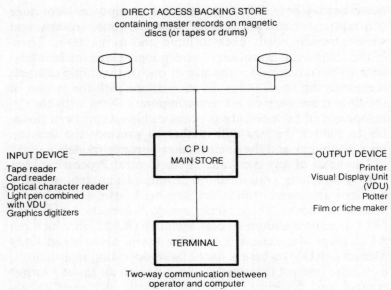

DIRECT ACCESS BACKING STORE
containing master records on magnetic
discs (or tapes or drums)

INPUT DEVICE

Tape reader
Card reader
Optical character reader
Light pen combined
with VDU
Graphics digitizers

C P U
MAIN STORE

OUTPUT DEVICE

Printer
Visual Display Unit
(VDU)
Plotter
Film or fiche maker

TERMINAL

Two-way communication between
operator and computer

Fig. 6.9

Data input. Current data in respect of work orders, job times, stores issued, etc. are input to the computer, using one of the input or communication devices.

Master file input. The master files contain records in respect of buildings, operatives, stores, etc. which are input into the computer in conjunction with movement data for updating.

Data processing. The processor carries out the operations defined in the program and transfers the results to the output area of the main memory.

Output. Finally the results of the data processing are transferred from the main memory to an output device which may be a printer for management reports or a visual display unit (VDU).

Control. This is exercised by the operator by means of a console unit or remote terminal. This permits two-way communication between the operator and the computer and enables the operator to stop or start the computer, to read in programs and data and to monitor the results.

Method of use

There are two basic methods of use – the 'batch system' whereby the user prepares data whch is processed at a later time by the computer, and 'on-line processing' whereby the user has a terminal connected to a remote computer for the immediate input and output of data. The major disadvantage of batch processing is the delay in detecting and correcting errors. Alternatively a microcomputer may be used as an 'intelligent terminal', allowing data from a central computer to be copied on-line to the microcomputer's memory and processed locally.

The computer may be self-contained or part of a communications network. The capability of computers to communicate with remote devices and of physically separate computers to communicate with each other is important in the design of many information systems. There are two broad classes of communications network – local area networks (LAN) and wide area networks (WAN). As the terms suggest, the difference lies in the distances between the devices. This dictates the type of linkage required, which for local area networks could be coaxial cable and for wide area networks the regular telephone network.

When physically separated computers are interconnected through communications facilities the configuration is called 'distributed computing'. The hierarchy of processors is illustrated in Fig. 6.10.

Software

Computer programs are referred to as 'software' and may be several times more expensive than the hardware. However, standard packages are available for the more common business routines. The first step in writing a program is to prepare a flow chart showing what has to be done in a logical sequence. This part is done by a systems analyst and then the programmer takes over and converts the flow chart into a series of instructions which are intelligible to the computer. The first computer languages, such as ALGOL (algorithm language), were developed by mathematicians and engineers

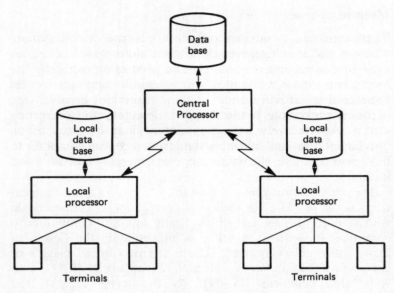

Fig. 6.10

and were not particularly suitable for business purposes. Then came FORTRAN (formula translation language) and COBOL (common business orientated language). More recently BASIC (beginners all-purpose symbolic instruction code) has been developed for the uninitiated. It is used for developing programs in a conversational mode in an on-line programming environment. As it is interpreted a line at a time any syntax errors are immediately brought to the notice of the user. This makes it slower than the other languages and not so easy to run sub-routines, e.g. in FORTRAN one can use a common pool of data and write sub-routines which dip into the pool as necessary to extract the relevant data.

Data files and bases

The basic unit is an 'entity' which is any type of thing or object about which data is collected. An 'attribute' is a characteristic of interest about an entity and each attribute is represented in storage by a data item. The data representation of an entity is termed a 'record' and a collection of records is called a 'file'.

The main types of files are:

1. Master file. This contains relatively permanent information about entities, e.g. description of buildings owned by the organisation, and is used as a source of reference data.
2. Transaction or work file. This is a collection of records describing activities and is developed as a result of processing data on work requested, planned and completed. It is used to give management feedback and to update the accumulated data in the master files.
3. Report file. This is created by extracting relevant data from the above files for the purpose of preparing reports.

The concept of a file as a set of related records resulted in physically separate files with 'owners' for each file. This has a number of administrative disadvantages in the form of restricted access to data, duplication of data and lack of data compatibility. The solution is a single, formally defined, centrally controlled database containing all related data files. The data records are organised and stored so as to be readily available to all interested sections of the organisation and to be capable of being developed in a compatible manner. The operational control of the database is obtained by means of a database management system (DBMS) which performs the function of defining, creating, revising and controlling the database. It provides facilities for retrieving data, generating reports, revising data definitions and updating data. Basically the DBMS controls the interaction between the database and programming users by means of application programs, and non-programming or 'ad hoc' users by means of the database query language facility.

Types of database

There are three main types of database: hierarchical, network and relational. In a hierarchical system records are linked in a tree-like structure or parent–child relationship. Network structures expand the hierarchical concept by allowing a given entity to have any number of superiors instead of only one. Relational databases are structured in a series of 'flat' sequential files with data repeated in different files (or tables) to form logical links which may be supplemented by pointers or index tables. Locating the data, therefore, relies upon the data in one table being related to that in another. Figure 6.11 illustrates this concept applied to housing maintenance assuming jobbing repairs are carried out by term contractors.

Whilst the relational structure is less efficient for processing large volumes of data it is more flexible than the hierarchical structure and can cope better with unplanned or *ad hoc* queries. Some databases incorporate a query/report utility, restructuring capabilities and data description languages to avoid much of the tedium and lack of user-friendliness associated with many of the earlier systems. Screen formatting, selection criteria and sort parameters are now commonplace

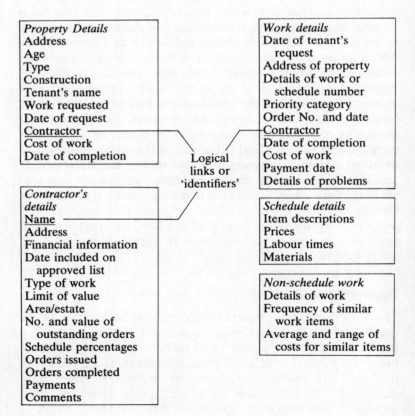

Fig. 6.11

and should be looked for when evaluating a system or package. Others include a data dictionary which permits the user to maintain a coherent assemblage of terminology. Irrespective of the size of the installation, back-up provisions are necessary to safeguard data from events such as fire, malicious damage or accident. Back-up and recovery of databases has been referred to as the Achilles heel of micros since, unlike their mainframe relations, they are often unable to dump data during the normal course of operation. It is essential, therefore, that diskettes or cassettes should be copied with appropriate frequency.

Word processors

Word processors are now used extensively in offices for the computer-assisted preparation of documents and correspondence. In essence they are no more than a microcomputer with CPU, visual display, floppy disk and a good quality printer. They may take the form of a single work station with a small amount of secondary storage, or multiple work stations sharing a single secondary storage device. Text is typed in the usual way and error correction can be performed very easily on the screen. Some of the more sophisticated systems allow the user to work on two documents at the same time and transfer blocks of text from one to the other. Others have an automatic spelling correction routine based on a predefined dictionary of words.

Uses

Computers have three main uses:

1. Mathematical. The carrying out of extremely complex mathematical calculations or 'number crunching' as it is sometimes referred to.
2. Commercial. The carrying out of relatively simple calculations on large amounts of data.
3. Process control. The use of microprocessors to monitor and control the operation of plant and equipment, e.g. optimising start and stop times for firing boilers.

The first two are extensions of the human mind whereas the

last one is taking over man's role as a machine minder. It is, however, the commercial uses to which the remainder of this chapter is devoted. It should be noted that in this respect the computer is doing little more than could be done manually but very much more quickly and with less risk of errors. It makes it possible to rearrange data to serve various management functions without having to employ an army of clerks for the purpose. Also, once the data has been committed to a computer file, it becomes possible to use the data in other people's files. Thus, one should be thinking in terms of an integrated system for the organisation as a whole rather than one which is limited to the maintenance department. Pettitt[11] described the objectives of the computerised system used by Peterborough Development Corporation as 'to record, store and process every item of repair and regularly produce a detailed breakdown and analysis of all maintenance work as an aid to achieving strict budgetary control and to retain and analyse the maintenance history and report on the frequencies of repetitive works of repair for the purpose of developing planned preventive maintenance programmes'. The system was originally installed for the use of the finance department and was adapted to give maintenance information by substituting maintenance repair codes for the finance department's codes.

More recent developments in computerised systems for maintenance management have been towards integrated systems which provide for the direct on-line entry of job information. However, before deciding on the type of hardware and software one should make a careful assessment of the organisational requirements in terms of the nature of the input data, the form of the output data and the number and location of access terminals and printers. The need to involve the people who will be using the system should not be overlooked. Initially it will call for a good deal of time and effort to produce the necessary databases which would normally include:

1. Property database: location, description, construction, age, use, etc.
2. Work specification: coded job descriptions with priority ratings
3. Estimating data: coded schedule of rates with items broken down into labour, materials and plant

4. Planned inspections: frequencies and check list of items to be inspected
5. Accounting data: codes for charging jobs to different cost centres
6. Contractors: details of contractors on approved list and details of contractual arrangements with term contractors
7. Direct labour: personnel files (see Data Protection Act, 1984 regarding processing of personal data by computer) and stores inventory
8. Historical data: details of various categories of maintenance and improvement jobs.

Usually an on-line system would be adopted to give immediate and up-to-date information in different forms to aid management decisions. Ideally the system should:

1. Provide for the continuous up-dating of the property database reflecting physical changes in the characteristics of the property and patterns of use.
2. Permit the direct input of details of new work to be carried out to the property including:

   (*a*) description of the work, expressed in a standard form
   (*b*) location of the work
   (*c*) trade involved for direct labour or name of contractor
   (*d*) priority rating
   (*e*) when access can be gained to the premises.

3. Schedule the jobs into an economic sequence taking into account:

   (*a*) location of the jobs
   (*b*) priority rating
   (*c*) age of the jobs
   (*d*) current availability of direct labour operatives or contractors.

4. Print out work orders either singly or in batches for particular buildings or areas.
5. Monitor the progress of individual jobs and report any that are falling behind the programme.
6. List all costs against work orders to provide up-to-date information on payments and commitments for budgetary control.

7. Check availability of materials in stores and issue reminders when stock is nearing re-order level.
8. Produce statistical information concerning:

   (a) number of jobs requested, completed, cancelled and in hand
   (b) average response times for dealing with user requests
   (c) analysis of backlog classified according to location, trades, contractors, etc., with reasons for holdups, e.g. no access, waiting for materials, special plant, instructions, etc.
   (d) performance measures for direct labour and contractors
   (e) recurring defects for inclusion in planned preventive maintenance programme and feedback to designers.

9. Provide the necessary financial data for accounting purposes including payroll procedures, bonus payments and contractors' accounts.
10. Build up a complete maintenance history of each building classified into elements and use areas with future projections of major renewals and repairs.

Clearly, the manual processing of the very detailed and fragmentary information on maintenance jobs to the degree of sophistication suggested would be extremely time-consuming, and such processing is only really feasible using a computer.[12] The main advantages that can be gained from processing such data by computer instead of manually are:

1. Provision of a more efficient maintenance service by resolving many of the day-to-day operational bottlenecks.
2. Overall reduction in maintenance costs which will more than compensate for the cost of installation.
3. Reduction of human errors and staff time on routine clerical processes.
4. Rapid availability of comprehensive operational and budgetary data.
5. Ability to rearrange data quickly to serve various purposes, e.g. a spot investigation of a particular type of defect.
6. Information may be accessed by other systems if there is a central database.
7. Better feedback to designers of defects related to particular forms of construction or materials.

Input **Processing** **Output**

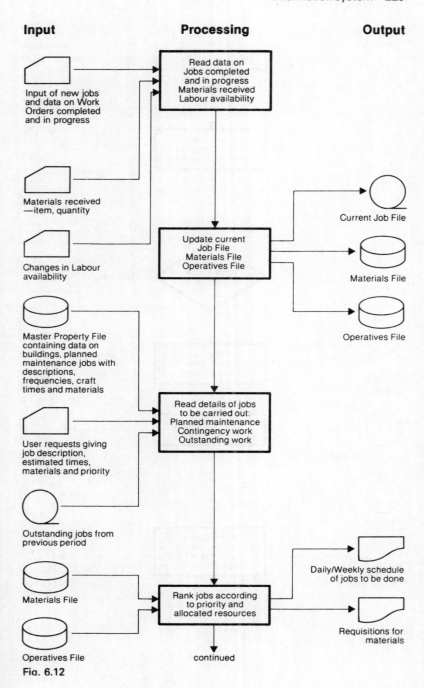

Fig. 6.12

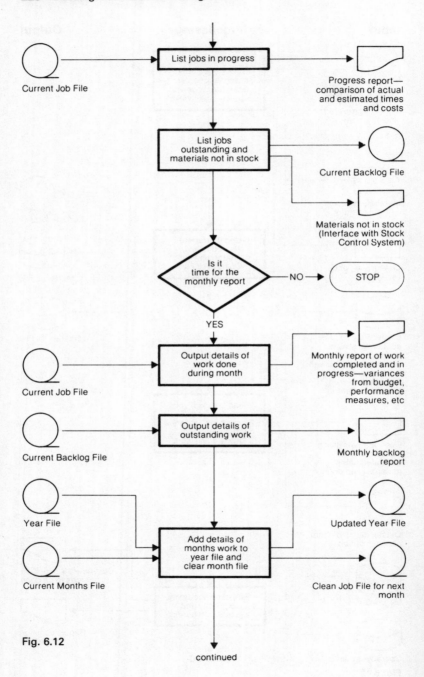

Fig. 6.12

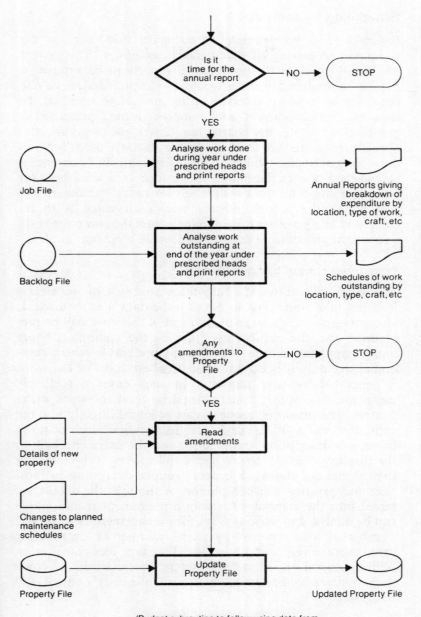

Fig. 6.12

Job File

Backlog File

Details of new property

Changes to planned maintenance schedules

Property File

Is it time for the annual report — NO → STOP

YES

Analyse work done during year under prescribed heads and print reports → Annual Reports giving breakdown of expenditure by location, type of work, craft, etc

Analyse work outstanding at end of the year under prescribed heads and print reports → Schedules of work outstanding by location, type, craft, etc

Any amendments to Property File — NO → STOP

YES

Read amendments

Update Property File → Updated Property File

(Budget subroutine to follow using data from updated Property File and analyses of previous years expenditure adjusted for changes in price levels and policies)

Scheduling by computer

Figure 6.12 is an example of a typical flowchart for the computer processing of maintenance schedules. The system illustrated provides for pre-planned reports to be printed at regular intervals but facilities would also be provided for *ad hoc* enquiries at irregular intervals using an on-line terminal. In some cases information is more comprehensible if presented in graphical form, e.g. bar charts, line charts, pie diagrams, etc. Spreadsheets are another modelling facility used for the development of plans, budgets and other analyses. In essence a spreadsheet defines the output in the form of a worksheet set out in columns and rows with labels on each. The statements that define the model indicate how each value is to be computed and provides for summing across the row to provide a row total and down a column to provide a column total.

Work order control by computer

The scheme illustrated is a fully integrated system and clearly it would take some time to develop the data files required. It also envisages that a large proportion of the work will be pre-planned and the sequencing done by the computer. More limited applications may be thought preferable where comprehensive data files do not exist or where most of the work is generated by user requests. In such cases a relatively inexpensive computer system could be used for work order control. The incoming requests can be keyed directly into the computer via a VDU keyboard using codes for location, trade, job description and priority and the entry checked on the display to avoid errors and duplications. The computer then prints out the work orders grouped according to location and priority. On completion of the jobs the details are keyed into the computer forming a permanent record which can be analysed in various ways for management control, e.g. number of jobs received by trade, number of outstanding jobs, average response times, etc. The data files can thus be built up over a period and the system later extended to cover such matters as bonus calculations, inventory control and budgeting.

References

1 Kochen, M. (1965) *Some Problems in Information Science.* New York: Scarecrow Press.
2 Institute of Cost and Works Accountants (1967) *Management Information Systems and the Computer. Part I: The Design of a Management Information System.* London: ICWA.
3 Robertson, D. (1973) 'Data collection and analysis.' London: National Building Maintenance Conference.
4 Ministry of Housing and Local Government (1964) *Report of the Working Party on the Costing of Management and Maintenance of Local Authority Housing.* HMSO.
5 *Building Maintenance Manual and Job Diary.* 2nd edition (1968) London: The Building Centre.
6 Department of the Environment (1970) 'Practice in property management.' R & D Bulletin, London.
7 Vickery, B. C. (1965) *On Retrieval System Theory.* London: Butterworth.
8 Building Research Establishment/Department of the Environment (1965) *A Study of Coding and Data Coordination for the Construction Industry.* HMSO.
9 Bushell, R. J. (Sept. 1970) 'Building planned preventive maintenance.' *Building Maintenance.* pp. 26–9.
10 Jarman, M. V. (1967) 'Selling maintenance to management – the use of proper costing.' London: Conference on Profitable Building Maintenance.
11 Pettitt, R. (1981) 'Computer aids to housing maintenance management.' HMSO.
12 LAMSAC (1980) *The Benefits of Using Computer Systems for Housing Maintenance and Repairs.* London: LAMSAC.

CHAPTER SEVEN

Organisation of maintenance department

Organisation theory

It is rare that one has the opportunity to plan the structure of an organisation 'de novo'. Usually the maintenance manager takes over an existing structure which has evolved to its present form over a number of years. Possibly the most important factor has been the personalities of the key members of staff. It is helpful, therefore, to understand the pressures and influences which have determined the nature of its development and of any constraints which may restrict future changes. There is no single structure which is equally appropriate for all organisations. The most appropriate is that which is best suited to meet the particular needs of the organisation.

A great deal has been written about the theory and practice of management and it is only possible in this book to give an outline of the main lines of thought on the subject. The early studies were concerned primarily with the individual worker then attention was turned to the organisation as a whole and the behaviour of groups of workers and more recently to the total system in which the organisation operates. F. W. Taylor, an American engineer, is generally credited with being the originator of scientific management. He realised the advantages to be gained by a systematic study of the way in which repetitive operations were carried out. He found that complicated jobs could be broken down into small units enabling workers to specialise in only a part of the whole job and thereby become more proficient and productive. He assumed that workers were motivated primarily by cash incentives and that the simplification of job method would not only benefit management by the resulting higher productivity but would also lead to higher pay for the workers and be mutually advantageous. F. Gilbreth, a building contractor, applied motion study to bricklaying and after careful study of the body movements involved in laying a brick devised ways of cutting out unnecessary movements so that a brick could be laid with only five motions instead of the eighteen required using the old method. H. Fayol, a French mining engineer, also applied Taylor's ideas and from them developed a set of general principles covering such aspects of management as division of work, authority, responsibility and discipline. However, others were beginning to criticise Taylor's basic notion that workers are motivated solely by cash incentives. Mary Parker Follet developed the Law of the Situation which saw management as a series of changing situations each dependent on and conditioned by the preceding situation. She argued that orders should not be given because 'I say so' but because the situation demands it. Thus managers are as much under orders as the workers in that both must obey the law of the situation. The simplistic views inherent in Taylorism were further challenged by the Hawthorn experiments. These were a series of studies carried out by Elton Mayo and others at the Hawthorn plant of Western Electric in America. Initially, a group of workers was tested to ascertain the effect of various levels of lighting on their rate of working. It was found that there was a general improvement as the lighting levels were progressively increased but that the rates continued to improve when the electricians only pretended to fit brighter bulbs. One explanation of this so-called Hawthorn effect was that the operators, being aware that they were being tested, displayed preconditioned responses to the situation, i.e. that the very act of monitoring the operators' performance influenced their behaviour. These studies provided a valuable insight into the behaviour of working groups and led to the 'human relations' movement which placed greater emphasis on the social environment of the workers. It was recognised that a work group had a particular personality which was distinct from that of its individual members. Also that informal groups, i.e. those formed by casual meetings, say, in the canteen, are just as important as the formal groups which have been selected and allocated to a joint task by management.

More recent studies have been concerned with management as a socio-technical system. Handy[1] uses the term cultures to describe the norms or values which determine the way in which an organisation is operated and structured. He classifies these cultures as:

1. The power culture which depends on a central power source. This is typical of small family businesses and entrepreneurial organisations which have few rules and procedures and which are highly dependent on the person or persons at the centre. They are quick to react to change.
2. The role culture in which roles and job descriptions are precisely specified. This is typical of local authorities and other bureaucratic organisations and is characterised by the laying down of job descriptions, limits of authority, procedures for communications, etc. It works well in a stable environment but is slow to react to change.
3. The task culture in which people and other resources are organised to achieve some set task. This is essentially a team culture in which individuals have a high degree of control over their work. It is useful for new ventures but difficult to exercise control except through allocation of resources.
4. The person culture in which the individual is the central point and the organisation exists merely to serve the interests of its members, e.g. barristers' chambers. Certain specialists in an organisation may also regard it as an opportunity to do their own thing and owe little allegiance to the organisation.

There may be different cultures in the same organisation, e.g. the main culture may be a role culture to deal with routine maintenance work but with a research and development section working on a task culture.

It is now popular to adopt what is called a systems approach. This stems from the work which was originally carried out by N. Wiener [2] who viewed an organisation as an adaptive system which is entirely dependent on measurement and correction through information feedback. The basic idea is illustrated in Fig. 7.1.

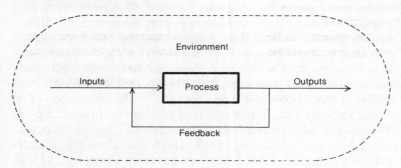

Fig. 7.1

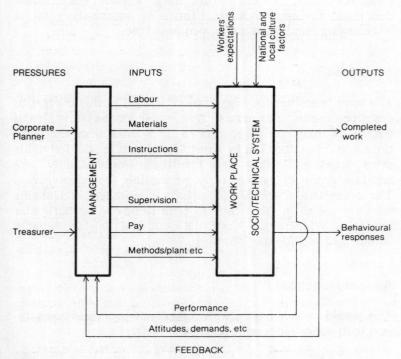

Fig. 7.2

The model may be extended to take into account all the social and economic pressures which affect the work group as illustrated in Fig. 7.2.

The move towards a more open systems approach has led to the contingency or situational approach. The contingency approach means that different environments require different organisations and that there is no one best type of organisation. For example, it means that job enrichment should be applied with the realisation that some operatives do not want their jobs enriched. Each person and situation is different and problems are created by the adoption of a single value system. Early evidence of this was provided by J. Woodward who showed that the form of organisation was related to the technology used which she classified as unit production (small batch), mass production (large batch) and process production. This has been followed by studies by Lawrence and Lorsch[3] who have shown that stable conditions favour the classical forms of organisation whilst in changing conditions the opposite is true.

Functions

The term 'maintenance organisation' is used in this context to describe the person or persons responsible for the planning and control of maintenance operations. In a small firm, the functions may be undertaken by a member of staff in addition to his other duties, while in a large firm there would usually be a separate group of people solely responsible for maintenance. The organisation would also include independent consultants who are called in from time to time to advise on particular problems. Whatever the scale, the basic functions are broadly similar and include the following.

Advisory function

This would involve liaison with owners and users and consultation with upper management on such matters as:

1. The standards to be maintained and the effect on user activities of deviations from these standards.

2. The relative merits of alternative maintenance policies and the extent to which it would be advantageous to employ operatives directly for executing the work.
3. Clarification of any constraints in relation to limits of expenditure, desirable cash flow patterns, acceptable delay times or restrictions on time and method of carrying out work.
4. Estimates of maintenance expenditure both long and short term, including, where appropriate, the cost of initially bringing up to the required standard and the possibility of phasing any such backlog over a period of years.
5. Provision of cost and other data to assist upper management in deciding whether to repair or renew.
6. Technical requirements for minor works involving alterations or small additions to the building; although not strictly maintenance, it is usual for the maintenance organisation to assume responsibility for this type of work.
7. Advice on the maintenance implications of designs for proposed new buildings; it is most important that the maintenance organisation should participate in the briefing of the designer and be given the opportunity to comment on the detailed design.

Organisational function

This may be in relation to the central administrative and supervisory system or to the execution system whether by direct labour or contract.

1. Central administrative system
These functions would be necessary and the associated costs incurred whether the work is ultimately let to outside contractors or undertaken by directly employed operatives.

(a) Defining the duties and responsibilities of administrators and supervisors and of their technical and clerical supporting staff.
(b) Establishing job relationships, patterns of accountability and paths of contact.
(c) Formulating standard procedures and operating instructions.
(d) Devising an appropriate information system and channels

of communication to ensure co-ordination and effective feedback for control purposes.

(e) Provision of suitable office accommodation and equipment.

2. Direct labour force

Where the work is undertaken by an independent contractor, he would assume the responsibilities in this area.

(a) Selection of foremen and definition of their duties and limits of authority; alternatively, the functions of the foremen may be considered in the context of the central administrative system.

(b) Engaging operatives, including payment, timekeeping procedures, administration of incentive schemes, provisions for safety, welfare and training.

(c) Purchasing materials and stores control procedures.

(d) Provision of plant either by purchase or hire.

(e) Arranging for the transport of labour, materials and plant to and from the site.

(f) Upkeep and maintenance of workshop and stores buildings.

3. Contract work

(a) Preparation of tender documents and selection of a contractor.

(b) Administration of the conditions of contract, including authorising and valuing variations and certifying for payment and satisfaction.

(c) Supervision of work to ensure compliance with terms of contract.

Control

The control functions are dependent on the timely receipt of accurate information relating to the state of the system. The control functions operate in the following areas:

1. Work input. Identifying the extent of work necessary to achieve the required standards within the constraints laid down. The processes involved would include planned inspections, appraisal of user requests and assignment of priorities.

2. Time of execution. Programming the workload so that the

carrying out of the work is timed in accordance with the needs of the user and the available labour force. This would normally involve the preparation of long-term strategic plans and of short-term tactical plans when the requirements are known with greater certainty.

3. Quality. Supervision of work during execution and by subsequent control inspections to detect latent defects.
4. Cost. Budgetary control system including estimating resource requirements in cost and performance terms for later comparison with actual cost and performance achieved.
5. Feedback. This is an inherent feature of all the control functions and would involve keeping such essential records as are necessary for the proper control of the operations.

Miscellaneous functions

In addition to the basic functions described above, the maintenance organisation may have responsibility for various other matters such as:

1. Safety and security, principally in relation to compliance with statutory fire precautions and the proper maintenance of fire fighting equipment.
2. Refuse disposal.
3. Cleaning.
4. Grounds, etc.

Structure

The structure of an organisation is usually represented by a chart showing the allocation of formal responsibilities and the linking mechanisms between the roles, i.e. the typical line and staff chart. This is usually backed up with a corporate plan setting out the general obligations and policies of the organisation and job specifications which outline the duties of the various members of staff and indicate to whom they are responsible and the limits of their authority. The types of structure appropriate to maintenance may be described as centralised (or functional) and decentralised (or territorial or

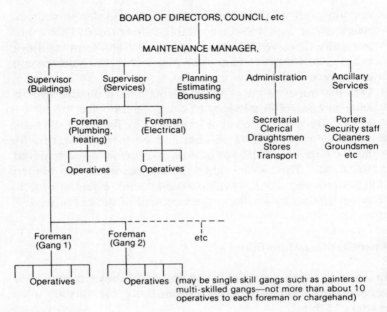

Fig. 7.3

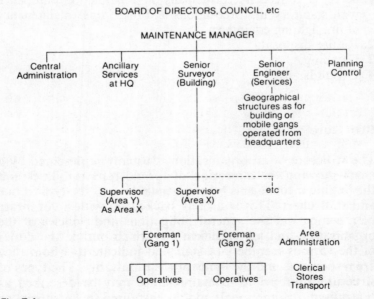

Fig. 7.4

geographical) as illustrated in Figs 7.3 and 7.4. The extent to which power is retained by the central organisation will depend upon the advantages to be gained from standardisation and the adoption of common procedures centrally administered and controlled.

The needs of different firms are so varied that there is clearly no one type of organisation which would be equally successful in all cases. A survey of a small sample of firms failed to reveal any uniformity of practice. In many cases, it was apparent that where the firm had expanded and acquired additional premises, the maintenance organisation merely increased in size without any change in its original character, in spite of the different circumstances under which it was operating. Much depended upon the personalities of the particular individuals concerned.

The size and structure of the central organisation will depend upon the following factors:

1. The volume of the workload will determine the staff time required for inspections, estimating requirements, preparation of drawings and technical documents, programming and control functions.
2. The nature and complexity of the work and whether predominantly building or engineering will determine the desirable qualifications of the supervisory staff.
3. The location and dispersal of the work will influence the travelling time and hence the number of supervisors required to maintain effective control.
4. The timing of the work and, in particular, the need for certain work to be habitually undertaken outside normal working hours may demand some duplication of supervisory staff to ensure continuous control.
5. The skill and reliability of operatives will determine the amount of information which they require and the frequency of visits to check progress and maintain quality control.
6. The method of executing the work, i.e. whether by direct labour or by contract, will have obvious repercussions on the organisational structure of the maintenance department.
7. The responsibility of the maintenance department for minor new works or miscellaneous services.
8. Building owners' or users' policies for different types of buildings.

In practice, one finds a wide range of structures ranging from the extremely informal arrangements common in small private firms, to the complex hierarchical systems adopted by large local authorities and government departments. A comparison of these extremes will serve to illustrate their relative merits and shortcomings.

In the small firm the proprietor may work alongside the men and discuss with them not only job information but also feelings, values and standards. In this type of organisation, the problems are usually only roughly categorised and the men have only a few loosely defined routines. The structure has few levels which are not clearly distinguished and co-ordination is achieved by the overlapping and intermingling of roles. It is extremely adaptable and well suited for dealing with the uncertainties of maintenance. However, it does require competent operatives and a stable working relationship between the members of the group. It is particularly advantageous where the firm employs highly skilled operatives who specialise in certain aspects of maintenance.

On the other hand, the organisation of a large local authority direct labour force does not permit the same degree of participation by the operatives. The work is usually divided into a number of small, clearly defined units which are managed by formalised routines. The relations between these units are laid down in further routines with detailed job descriptions and rules and procedures for each man. Co-ordination between the units is achieved by a hierarchy of positions with the powers of each clearly defined. The principal defect is the lack of flexibility and inability to deal with conditions which were not foreseen when framing the rules and procedures. It works best where there is a repetition of the same problem for which the method of working can be clearly defined. It does not, therefore, require the same degree of expertise on the part of the operative and there is a restricted scope for the exercise of discretion.

Also, the large complex organisations demand a level of management expertise not readily available in this area. The Emmerson Report[4] noted the lack of people with appropriate management skills throughout the construction industry and that those who have the necessary qualifications are attracted more to new work than to maintenance. This point was emphasised in the Woodbine Parish Report in relation to maintenance supervisors in the hospital service. Of the 240 or so building supervisors employed by the service in 1969, 85 held no technical qualifications and only 63 had a Higher National Certificate or higher qualification. It was reported by the hospital authorities that those with the higher qualifications did in fact achieve better standards of maintenance and improved organisation.

A report by the Construction Industry Training Board [5] drew attention to the serious weaknesses in the management structures of small and medium-sized building firms. To combat such problems, the report recommended broadening the educational preparation for supervisory and management roles. However, it would seem that the situation can be improved only by recognition of the importance of maintenance and reassessment of the status and rewards due to persons responsible for the organisation of this work. In the case of local authorities, it has been noted [6] that the organisation of maintenance may form part of the duties of the engineer, the architect or the housing manager, but only rarely is there a building manager with sole responsibility for planning the execution of this work. Because of the routine nature of such maintenance work, authorities are inclined to leave effective control in the hands of officers who regard it as secondary to their main duties and are not equipped to consider the longer-term issues involved.

Management at every level develops its own policy within its own sphere of responsibility and scope for decision making. Broadly, policy is to do with the setting of objectives and the means of achieving them. Decisions at the different levels in the organisation must be compatible and clearly understood and implemented by all who are directly concerned; others who may be affected but who are not directly implicated must be informed. Clearly, success is dependent upon the ability to set and achieve the best objectives at all levels. This has been described as 'management by objectives'. It should not be overlooked that people work more effectively together if their individual objectives are parallel and do not diverge. A major part of labour relations is to do with keeping the objectives of all concerned sufficiently close to ensure that the objectives of each are achieved by the joint efforts of all. 'Management by exception' is another important concept in this context. It means that by having established objectives and the means of achieving them attention can be concentrated on deviations or 'exceptions' from the main plan. This involves providing adequate means for monitoring performance against targets set, so that exceptions can be identified sufficiently early for corrective action to be taken.

Direct labour or contract

The choice between employing tradesmen directly to execute maintenance work or engaging an independent contractor for that purpose should be decided according to which offers the greater advantage in terms of cost, quality and convenience. Usually a large organisation will use both methods, the problem then being to decide the amount and type of work which should be carried out by each. Potentially, direct labour is more economic than contract work by at least the profit margin included by the contractor in his tender. However, achieving this saving demands an equivalence of management skills so that the performance of the directly employed operative is not inferior to that of the operative employed by the contractor. There are other factors to consider besides cost and a decision must rest on an analysis of the particular advantages and disadvantages in the context of the needs of the organisation.

Main advantages derived from directly employed labour

1. Full control of the allocation of work to operatives, resulting in a quicker response to emergencies and greater flexibility in terms of the work done and the times worked. Where the work is carried out by a contractor the instructions must pass through the contractor who may be unable to withdraw men from other jobs as quickly as the situation demands. Losses flowing from the interruption of the firm's activities during these periods of delay should be taken into account when comparing the costs of direct labour and contract.
2. Directly employed operatives acquire an intimate knowledge of the buildings and become familiar with the user's requirements and any constraints on the execution of the work.

This simplifies the communication of job information in that the operative has a background knowledge of the situation and is able to understand what is required more easily and with less risk of misunderstandings. It also engenders a better working relationship between the user departments and the maintenance department.

3. Better control of quality through the employment of men of known and tested ability and through direct supervision. The directly employed operative will have a greater sense of identity with the firm and will see maintenance as a continuing process in which defective work will only create greater problems at a later date. There is thus a greater incentive for the directly employed operative to do the work properly the first time. However, whether or not this advantage is realised will depend upon the policy of the user organisation. In a report on local authority incentive schemes it was noted that it was the policy of a number of councils to give employment to men who were otherwise unemployable. While it might be socially advantageous to provide welfare in this way, it results in maintenance costs being higher than they need be and prevents the comparison of direct labour and contract on the same basis.

It is important to consider the total cost of maintenance over a period of time rather than marginal savings on an individual job. Thus there is nothing to be gained in the long run by skimping the preparatory work for external painting in order to reduce costs. The result will be a shortening of the painting frequency and a need for more extensive work at the next repainting period. For such work it is generally thought that direct labour ensures a higher quality and a longer lasting job.

4. By employing operatives directly it is possible to gain a first-hand knowledge of the factors which influence output and thereby to develop more effective cost control procedures linking planning and execution. It is also possible to carry out controlled experiments with different methods in order to achieve the same standard at a lower cost. Usually, where work is let out on contract, only the lump sum price is recorded for feedback purposes. This is of little value for estimating and programming future work for which labour hours would be more useful.

5. Delay is not incurred in inviting tenders or negotiating with contractors, and the procedures for initiating the execution of work are simplified.
6. Where the work involves a security risk it is advantageous to have knowledge of the background of the operatives.

Disadvantages attributed to direct labour

1. There is a lack of specialisation in terms of labour skills and plant. However, this criticism could be applied equally to the small building firms which execute the major part of maintenance work. Many small property owners make great use of the jobbing builder who employs a few handymen and calls upon self-employed tradesmen where necessary. On the other hand, the larger direct labour forces are demonstrably better equipped than most of the very small, severely under-capitalised independent contracting firms. The assertion is really only valid in respect of certain types of work, e.g. remedial treatment to timber, painting, cleaning, etc., in which specialist firms have developed a particular expertise.
2. There is a feeling in some quarters that direct labour is more expensive than contract work. Indeed, this is the most common complaint levied against direct labour forces employed by public bodies. It was for this reason that the Local Government Planning and Land Act, 1980, required local authority direct labour organisations to tender in competition with contractors for works over a certain value and to show a prescribed rate of return on capital involved.

   However, comparing tenders is not a valid way of measuring either the efficiency or the usefulness of a direct labour force for the following reasons:

   (a) The submission of the lowest tender does not automatically indicate that the contractor submitting that tender is the most efficient of those tendering. A low tender may be the result of inaccurate estimating or of a shortage of work in the locality, or it may be set deliberately low with the intention of cutting the quality. The true measure of efficiency, as far as the contractor is concerned, is the margin of profit; clearly there is no advantage in submitting a low tender if the result is a net loss. Direct labour forces do not make a profit and therefore must measure efficiency in other ways. The only true comparison would be of basic costs of labour and material expended on similar jobs, but as no two jobs are identical the result would be distorted by differences in the nature of the work and the conditions under which it has to be executed.

(b) A contractor is free to bid for any work available and, not being restricted to a particular client, may select that which presents the fewest difficulties. On the other hand, the direct labour force is confined in its operations and is frequently placed in the position of having to carry out work which by its smallness or complexity is not attractive to contractors. In such cases either it is not possible to find a contractor who is willing to do the work or the price quoted is excessively high.

(c) Continuity of work within an overall plan is one of the biggest economic advantages to be gained from the use of a direct labour force. It is, therefore, self-defeating to require it to justify its efficiency on each and every individual project by tendering in open competition with outside contractors. The result would be to introduce uncertainty in the planning which could disrupt the even flow of work programmed for the force, resulting in idle time and underutilisation of plant and inhibiting the forward bulk-purchasing of materials.

(d) There are inevitably many small, time-consuming jobs which usually, either because of size or urgency, fall to the lot of the direct labour force. In a hospital memorandum[7] it was stated that up to 50 per cent of the total maintenance budget was absorbed in dealing with minor breakdowns and *ad hoc* requests. Although it was anticipated that this proportion would be considerably reduced by the introduction of a planned maintenance system, the need for a 'fire brigade' service could not be eliminated altogether.

(e) The overall efficiency of a direct labour force cannot be judged on the basis of the result of an individual competition. A direct labour force is not able to balance gains on one contract against losses on another and the single criterion of cost can be misleading unless it takes into account all the benefits.

3. A further disadvantage which is sometimes pointed out is that the operatives may be at times underemployed. This may be inevitable if the circumstances demand an extremely short response time. In such cases, efficiency should be measured by the average time between failure and rectification as well as the cost of providing the required service. However, this sort of situation would merit close investigation to discover whether the emergency calls can be reduced by increasing the level of preventive maintenance. Filling the time with work of doubtful value is, of course, something which must be avoided.

4. It should not be overlooked that the use of direct labour necessitates the provision of supporting facilities in the form of building space for offices, stores and workshops as well as additional administrative and clerical staff to deal with work control, incentive schemes, labour relations, payroll and accounting services.

5. A further point to take into consideration is that the financial and other risks are borne solely by the organisation employing direct labour. Where work is let to outside contractors on a lump sum basis the extra costs occasioned by unforeseen circumstances are usually the contractor's responsibility and, except as provided in the contract, cannot be passed on to the client. Also the contract normally contains safeguards which entitle the client organisation to redress if the work is not carried out in accordance with instructions or is not completed on time. However, many of the smaller jobs are let on a cost reimbursement basis and this throws the financial risk on to the client who is at the disadvantage of not being able to control directly the job method.

Factors influencing choice

The aforementioned advantages and disadvantages should be weighed in relation to the following factors.

1. Nature of work

The programmed work should be divided according to the skills required for its execution, i.e. traditional craft skills, specialist skills and relatively unskilled or semi-skilled work.

(*a*) Work requiring conventional craft skills would normally be the main commitment of a direct labour force. In certain cases the firm may employ skilled craftsmen in other capacities, e.g. shop fitters, who are able to cope with the day-to-day requirements of building or services maintenance.

(*b*) Specialist work is usually let to outside contractors since there would rarely be sufficient work to provide continuity for a permanent labour force. Also a specialist firm would have a better knowledge of the characteristics of the component and possess the necessary tools and spare parts to effect a repair more quickly. In some cases the appointment of a specialist firm may be dictated by safety considerations, e.g. electrical installations, lifts, etc.

(*c*) For relatively unskilled work it may be thought more convenient to employ a handyman or to make the caretaker or user responsible.

2. Volume of work

The total amount of work in each of the above categories for each trade (or compatible combination of trades) should be assessed and annual and seasonal variations identified. From this it will be possible to determine:

(*a*) Whether or not the amount of work of one type will provide full and continuous employment for at least one person in each craft group. The number of operatives can be obtained from the estimated costs included in the long-term programme by assuming an average ratio between total cost and the manhours included therein.

(*b*) The work which should be let to outside contractors to deal with peaks of activity beyond the capacity of the direct labour force.

For this the workload should be analysed into:

(i) Routine day-to-day and frequently recurring minor works.
(ii) Major works divided into 'cyclic' and 'infrequent'.

An outline programme for the budget period taking into account any restrictions on the timing of the work imposed by the user of the building will reveal the peak periods when outside assistance will be required.

3. Response time

The workload should be analysed according to the degree of urgency of the work sections so that an assessment can be made of the response time necessary to avoid consequential losses or inconvenience. A breakdown into three categories would be adequate.

(a) Emergency work – the breakdown of vital services or fabric defects involving risk to the occupants or contents. Clearly this work must be dealt with promptly, preferably on the same day. However, it is rare that breakdowns calling for emergency repairs are completely unpredictable. In most cases the contributory conditions are discernible some time before the actual failure. The distinction lies in the length of the period between the time when the failure conditions are just capable of being identified and the time of failure. For instance, an unlagged exposed water pipe is clearly at risk and precautionary measures can be taken long before there is the probability of frost action. On the other hand, an exceptionally heavy fall of snow may quickly build up to a dangerous thickness on a flat roof and call for prompt action to remove it. Procedures should be laid down for initiating and executing such work outside normal working hours. In many cases, to be effective, speed is essential and this would favour direct labour or special arrangements with a contractor.

(b) Urgent work – defects which while not a positive danger cause some inconvenience to the user and may develop into more serious faults if left unattended. An acceptable response time would be one week and this again would tend to favour the quicker initiation of work possible with direct labour.

(c) Normal work – work which while necessary and desirable does not affect the immediate user of the building and which may be phased in with the annual programme. Whether it is carried out by direct labour or contract will depend very largely on its magnitude.

4. Location

Different views have been expressed on the relative merits of direct labour and contract for the maintenance of buildings in remote areas. In some cases the circumstances may favour the direct employment of labour in areas where the use of contract labour is difficult by reason of remoteness. In other cases it may be found more convenient to employ local contractors on an 'on call' basis, the object being to avoid excessive travelling time caused by calling direct labour from a central unit. This difficulty could be overcome by decentralising the organisation of the direct labour force if the individual buildings or groups of buildings provide a continuous workload for a resident maintenance team. The central workshop would provide a pool from which specialist staff and equipment could be sent to peripheral units should the occasion demand: a facility not available to the small local builder.

5. Quality

The importance of maintaining high standards must be considered. In the case of a departmental store it was the opinion that maintenance expenditure was all part of the business of selling and that everything must be done quickly and well to ensure that the store functioned at the highest possible level. It was considered that a high standard of maintenance, by attracting customers, was a source of profit and that therefore cost was of less significance than achievement. The surest way of obtaining such standards is by the direct employment of highly skilled operatives who can be more closely supervised than contract labour.

6. Security

The need for special security precautions may be the dominant factor overriding the probability of any marginal cost advantage. This might preclude the presence in the building of a contractor's casual labour who have no responsibility to the firm.

7. Availability of space

The employment of direct labour necessitates the provision of supporting facilities such as stores and possibly workshops. The question arises therefore as to whether it is physically possible to provide the space or whether an alternative use of the available space would be more profitable.

8. Market conditions

The decision to employ a direct labour force is essentially a long-term one and therefore an assessment should be made of the probable future differences in cost as well as those which exist at the present time. It is clear that the prices charged by contractors will be influenced to a very much greater degree by market conditions than will the costs of direct labour. The most variable component of the contractor's price is his profit margin and this is completely absent from direct labour costs. Also it is probable that in times of severe labour shortage the greater permanence and security of direct employment will be regarded by the older and more mature tradesmen as being preferable to higher wages for an uncertain period.

Market conditions vary according to time and place. The time-related changes may be long-term and progressive or short-term and fluctuating. The short-term changes may be of a predictable seasonal character or unpredictable and isolated. Different locations may follow a broad regional pattern and depart from the general trend in a consistent way or may be highly individual. The inconsistencies may arise from various causes: for example, the closure of a major industrial plant may result in a local depression with consequential effects on maintenance and other prices. Another factor which impinges directly on maintenance is the volume of new construction in the locality. Most small and medium-sized builders carry out new construction in addition to maintenance and usually prefer new work because it is more straightforward and offers potentially higher profits. Any curtailment of the new construction programme either as a result of the general economic climate or deliberate government policy will divert these firms to the maintenance and improvement market, thereby increasing the competition in this field with a likelihood of lower prices.

9. Cash flow

A further factor is that the larger part of maintenance expenditure is attributable to labour, and where operatives are employed directly wages must be paid at weekly intervals. Payment for contract work is not made until after completion of smaller jobs, or at monthly intervals or predetermined stages for larger jobs. In addition, it is normal to hold an agreed percentage of the amount due to the contractor until the end of the defects liability period. Of course, the cost of providing working capital should be included in the contractor's tender, but this may well be less than the profit which could be obtained from an alternative use of the finance by the client organisation.

Consideration of the foregoing factors will enable work to be apportioned between direct labour and contract in a manner which will best serve the interests of the organisation.

There are various compromise solutions designed to blend the best features of direct labour and contract. For example, an arrangement can be made with a firm of contractors to keep a mixed gang continuously on the site under a resident foreman. The instructions would be given directly to the foreman by the maintenance supervisor and the work paid for on a schedule or cost-plus basis. The main advantages obtained from this system are that the men are permanently on the site although the number can be varied according to the workload. The client organisation avoids the responsibility for employing labour in regard to wages, pensions, holiday payments, trade union negotiations, etc., and any special equipment can be obtained quickly from the contractor. Also the contractor can buy common materials in bulk and obtain more favourable terms from manufacturers and merchants. It is probable that the continuity of work resulting from this arrangement would encourage the contractor to quote lower prices.

10. Total costs

Where an organisation uses both direct labour and contractors the object should be to minimise the total costs of executing maintenance work as represented in the following equation:

$$T = (C + S) + (L + M + P + S) + A$$

where

T	=	total costs
C	=	contractors' costs
S	=	supervision costs
L	=	labour costs
M	=	materials costs
P	=	plant costs
A	=	administrative costs

L, M, P = direct labour

Comparison of directly employed labour and contract

A report of a DOE Working Party on Direct Labour Organisations and a discussion document on Direct Works Undertakings Accounting (Maintenance) by the Chartered Institute of Public Finance and Accountancy were published about the same time in 1978 and made similar proposals in regard to local authority direct labour forces. They suggested two ways of comparing the efficiency of direct labour with outside contractors:

1. The valuation of jobs on the basis of a Schedule of Rates derived from competitive quotations from private contractors.
2. The charging of jobs by a system of Target Hours on the basis of a competitive cost per hour derived from contractors' quotes in terms of both time and price for a typical job.

The Schedule of Rates method involves preparing a list of the most frequently occurring items with all-in prices for each item obtained from competitive quotations by contractors. The schedule rates are used for pricing work carried out by the direct labour force and the result compared with the actual cost of resources expended. In order to avoid bias the contractors should be assured of a representative share of the work otherwise the prices may be unduly high. Also, difficulties were reported in the case of one local authority which required a much more detailed schedule for bonus purposes than was required by the contractors for pricing.

The Target Hours method demands the existence of a comprehensive work-study-based bonus scheme. The procedures are:

1. Work carried out each week is expressed as a total number of productive standard hours (PSHs) which is calculated from the hours shown on the operatives' time sheets for actual work done (non-productive hours are ignored) and the labour constants used for the incentive scheme.
2. The costs of labour, materials, transport and overheads are aggregated and the total divided by the number of PSHs to give the total cost per PSH for the direct labour force.
3. Private contractors are invited to tender for a reasonable share of the work by quoting prices for schedule items covering the normal run of maintenance jobs required.
4. The contractor's rates are then used to price out a model which reflects the frequency of the individual items of repair over a given period.
5. The total number of PSHs contained in the model is found by applying the work-study-based data.
6. The total contractor's cost for the model is then divided by the total number of PSHs represented by the model to give the contractor's cost per PSH.
7. The contractor's cost per PSH may then be compared with that of the direct labour organisation.

Clearly, it is important that the mix of items in the model should be truly representative and that the contractors should base their quotes on the assumption that they will receive a fair share of the work. In addition it would be necessary to compare the quality of the work done and matters such as the time taken to respond to tenants' requests.

Also the reports recommended that the relationship between the direct labour organisation and the client authority should match as closely as possible those between a private contractor and the authority. They recommended that the direct works organisation should be placed in a trading situation and charge the local authority on the basis of the rates prevailing at the time for similar work recently won in competition. This would involve setting up a reserve fund into which any profits can be paid and from which losses can be defrayed with provision for eventual profits to be used for the benefit of ratepayers and suitable action in the event of continuing losses. However, it should be recognised that local authority direct labour organisations provide additional services, e.g. source of practical construction knowledge, involvement in policy matters and corporate planning and social functions in providing employment for apprentices.

Work order system

The key document used for controlling work input is the Work Order (WO) or Job Card. It provides a medium for the recording of labour hours and materials which can subsequently be analysed for various management purposes. There are in use many different types of forms of varying degrees of complexity, but the basic information they should give is:

1. Number and date of issue.
2. Codes to indicate location and other classification categories into which costs are required to be analysed.
3. Priority for non-schedule work. The priority will depend on the effect of the state of disrepair on the user of the building and in particular on safety and health considerations. It is difficult to make fine distinctions, but in practice three categories should be sufficient:

Priority I   Emergency work to be carried out on the same day.

     II   Urgent work which while not constituting an

immediate danger should be carried out within a week of notification.

III Normal work which would account for the major part of the budget allowance for contingency maintenance and may be programmed on the same basis as the scheduled work.

4. Description of work. As the work orders are the main source of information on costs for subsequent estimates the descriptions should be standardised, i.e. identical descriptions should be used for repetitive jobs. Usually the descriptions are too vague to identify precisely the work content, the presumption being that the details will be transmitted verbally on site. This not only leads to delays on site, but also renders the work order virtually useless for estimating purposes. A schedule of standard descriptions would be of considerable value, in that they could be coded to save time in making entries, and also it would facilitate the storage, retrieval and processing of cost data.

However, standard descriptions may not indicate the particular conditions under which the work has to be carried out and therefore, if an unusual method or sequence of operations has to be adopted, these should be noted so that future comparisons are not distorted.

5. Estimated labour hours. The estimated labour content expressed in man hours or man days should be entered on the order or, where a bonus scheme is in operation, the target time. Quite apart from the control value of comparing actual against estimated times, it focuses attention on the job method and the need for proper programming.

6. Materials. The quantities of each material should be stated in sufficient detail for stores requisitioning or purchase.

7. Actual labour hours. The actual labour hours worked on the job are required for cost control and bonusing. However, as they will form the basis for future estimates, a distinction should be made between productive and non-productive time and, where the latter is excessive, the reasons why this was so should be noted. It is also necessary to distinguish between emergency work and that of a lower priority, since this will govern the extent to which jobs can be grouped to reduce travelling time and hence costs. This will make the

additional cost of providing an emergency service explicit for comparison with the benefits obtained by way of extra safety and convenience.

All work carried out should be covered by an appropriate work order. A typical system for dealing with user requests would be as illustrated in Fig. 7.5.

The alternative adopted at Stage 3 will depend upon the status of the foremen in the organisation. For this type of work the foreman is usually in the best position to estimate the output of his particular gang of men. On the other hand, an independent estimate provides a better and more consistent check on efficiency. In some cases, the foreman may indicate the scope of the work and the materials required and a work study officer determines the standard hours by reference to elemental data.

The process also assumes that at Stage 4 it is possible to determine accurately the scope of the work, whereas sometimes preliminary exploratory work must be carried out before this can be done. In such cases, the job descriptions tend to be very loose so as to cover every possible contingency and as a result are useless for cost control and future estimating purposes. It is important, therefore, that the detailed job description should be entered on the work order as soon as it is known and the estimated labour and material requirements altered accordingly.

It is also apparent that where there is a degree of uncertainty a fair amount of discretion must be given to the foreman to arrange short-term programmes according to actual conditions prevailing at the time. Rigid adherence to programmed dates may not only prove impossible but, if attempted, lead to inadequate work.

Figure 7.6 shows in diagrammatic form an integrated system of work input control and the related procedures.

Stock control

The extent to which materials are purchased and ·stored in advance of use will depend upon the needs of the maintenance organisation, the ease with which materials can be purchased from local merchants and the overall cost consequences. The

CONTINGENCY SYSTEM (USER REQUESTS)

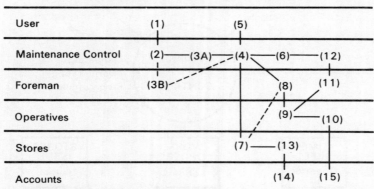

Fig. 7.5. Contingency system for user requests

1. User issues request to maintenance control centre describing defect
2. Request examined by maintenance control to check whether necessary and, if so, priority
3A. Preview of defect by building supervisor to determine scope of work and job method, or
3B. Alternatively, preview by trade foreman who will be responsible for supervising execution
4. Preparation of work order entering date, location, number, job description and estimated labour hours and materials
5. Copy of work order to user department for information
6. Copy of work order filed with open jobs
7. Copy of work order to stores to check availability of materials or to order out-of-stock materials
8. Copy of work order to foreman to arrange with stores for delivery of materials to site at the programmed time
9. Instructions to operatives on method of execution of work
10. Operatives complete time sheets giving details of hours worked, overtime and non-productive time
11. Foreman enters times for individual jobs on work orders and checks weekly total against time sheet
12. Work orders for completed jobs extracted from job file. (At the same time, progress on incomplete jobs checked and jobs not started investigated)
13. Stores control procedures
14, 15. Payroll and accounting procedures

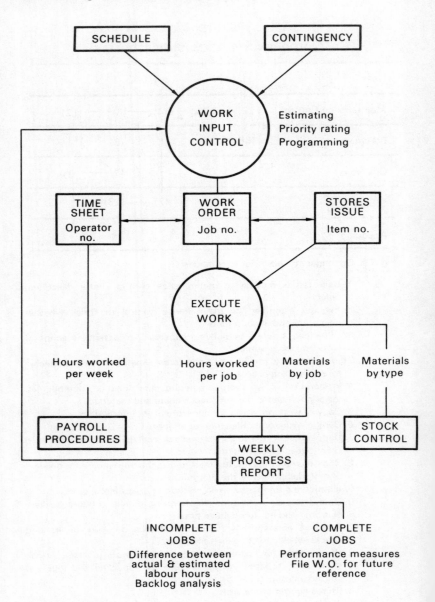

Fig. 7.6. Work input control

objects of stockholding are, firstly, to enable materials to be purchased in economic quantities and, secondly, to ensure that materials are available as and when required for the execution of urgent repairs.

There are certain advantages to be gained from buying materials in large quantities, such as lower costs for ordering, handling and transportation and perhaps a quantities discount. On the other hand there are costs associated with the holding of stocks, including the use and upkeep of a storeroom, attendant labour, risk of deterioration and obsolescence of the stock, insurance and, by no means least, the interest on capital tied up in stock. It is therefore a matter of determining the 'economic batch quantity' which will minimise the total costs as illustrated in Fig. 7.7.

The economic quantity for ordering may be calculated using the following equation:

$$Q = \sqrt{\frac{2cd}{sp}}$$

where Q = economic batch quantity
p = item cost

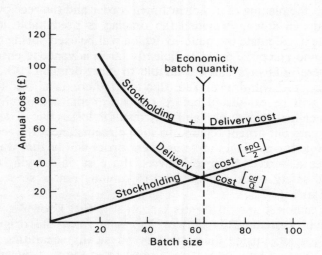

Fig. 7.7. Calculation of economic batch quantity

$d =$ annual demand for item
$s =$ stockholding cost expressed as a fraction of stock value
$c =$ delivery cost per batch

For example, assuming the annual demand is 200 items, the cost price per item £5, the stockholding cost assessed as 20 per cent of the stock value and delivery charge £10 per batch; the economic batch quantity would be:

$$Q = \sqrt{\frac{2 \times 10 \times 200}{0.2 \times 5}}$$
$$= 63.24 \quad \text{say} \quad \underline{65 \text{ items}}$$

However, this assumes that the only variables are delivery and stockholding costs. If there is a discount for quantity it becomes necessarv to compare the saving resulting from the quantity discount with the extra cost of holding the additional stock.

Where the rate at which items are withdrawn from the store is constant and replenishments can be made immediately, i.e. the 'lead time' is zero, there is no need to hold a buffer stock for contingencies and replacements should be ordered immediately the stock falls to zero. Usually, however, there is a timelag between the placing of a replenishment order and the receipt of the items in stock. Provided the timelag is predictable it is possible to calculate the quantity which will be used during this period and place the order sufficiently far in advance to ensure that there is always stock available to meet demand. Where, however, the withdrawal rate fluctuates there is a risk that stocks will be exhausted before replenishments are received. This is called a 'stock out' and may result in losses due to delay in carrying out urgent repairs. In such circumstances the economic time for placing a replenishment order will be that which achieves a correct balance between the cost of carrying additional safety stock and the cost of running out of stock, as shown in Fig. 7.8.

The simplest system of stock control is that known as the 'two bin' system. This derives its name from the original method of organising the system by physically separating the stock into two bins. Stock was drawn from the first bin until empty and then replenishment stock ordered. When the replen-

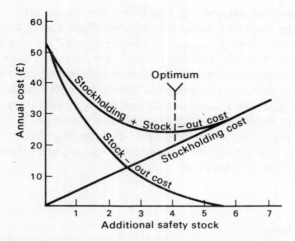

Fig. 7.8. Calculation of optimum stock-replenishment levels

ishment items were received the second bin was topped up and the remainder of the items placed in the first bin from which items were issued. This is also known as the re-order level system and may be administered by noting receipts and withdrawals in an inventory, by hand posting or by machine or electronic computer, and ordering replacements when the stock falls below a predetermined level. An alternative system known as the 'periodic review system' may be used where a series of items are obtained from a single supplier and it is desired to gain the maximum quantity discount. This involves placing orders at regular intervals for quantities calculated to bring the total stock up to some predetermined level.

All withdrawals should be on production of a properly authorised requisition which should state clearly the description of the material, the quantity required and the work order number. This can then be used for the purposes of cost accounting and stock control. The only exceptions would be items too small to warrant being controlled in this way, i.e. low-cost items used in large quantities (nails, screws, glass-paper, etc.), which could be issued on demand and charged as a percentage on the value of work done.

A physical check should be made of the items in stock at periodic intervals in order to detect errors in the inventory records. If there are serious unexplained shortages an investigation should be carried out and corrective action taken. At the same time the stock should be examined for obsolescence and deterioration and if necessary disposed of. The risk of this happening will be lessened if steps are taken to achieve the maximum degree of standardisation of components and fittings. Wherever possible replacements should be with standard parts even though some additional minor work is necessary. In the case of proposed new buildings the brief given to the designer should point out the advantages of reducing the variety of finishings and fittings in the context of the overall maintenance policy of the organisation.

References

[1] Handy, C. B. *Understanding Organizations*. Penguin.
[2] Wiener, N. (1948) *Cybernetics*. MIT.
[3] Lawrence and Lorsch (1967) *Organization and Environment*. Harvard University Press.
[4] Ministry of Works (1962) *Survey of Problems before the Construction Industry*. Report by Sir Harold Emmerson. HMSO.
[5] Construction Industry Training Board (1975) *Management/Technician Roles in the Construction Industry*.
[6] Layton, E. (1961) *Building by Local Authorities*. Report of an inquiry by the Royal Institute of Public Administration. London: Allen & Unwin.
[7] Department of Health and Social Security (1964) *Maintenance of Buildings, Plant and Equipment*. Hospital Technical Memorandum 12. HMSO.
[8] Luke, R. 'Emergency services.' (1969) London: Conference on Building Maintenance.

CHAPTER EIGHT

Labour relations and productivity

The labour force

Maintenance is a labour-intensive activity and therefore the greatest economies are likely to flow from measures which improve labour productivity. Studies[1] have shown that in the case of painting the labour content may be as high as 85 per cent of prime cost and for general repair work approximately 65 per cent of prime cost. The small scale of much of the work and the restricted conditions under which it is executed preclude the extensive use of mechanical aids. Although portable hand power tools could be used much more widely than at present, it is unlikely that this in itself would have more than a marginal effect on total costs. As noted in a report on the construction industry,[2] the small builder involved in this type of work is usually reluctant to meet the initial cost of even relatively simple new methods and techniques. As far as materials are concerned, the main factor is to ensure that they are available on site when required, so that time is not wasted awaiting delivery.

The principal consideration is the effective deployment of labour within an overall plan and for this purpose it is necessary to examine the characteristics of the maintenance labour force and the extent to which incentive schemes are likely to prove beneficial.

Operative skills

About 40 per cent of all construction operatives are employed on maintenance and repair, in the main by small local firms. This does not include operatives directly employed in commerce and industry, and taking these into account it is probable that the total manpower employed on maintenance and repair is almost as numerous as that employed on new construction.

The trades employed in repairs and maintenance are effectively the same as the broadly based construction trades and many small contractors use the same operatives for both types of work. The five main trades (carpenter, bricklayer, plasterer, plumber, painter) account for 80 per cent of the maintenance and repair labour force, the predominant trade being painters, who comprise some 30 per cent of the total. Indeed about 75 per cent of all painters and decorators are employed on maintenance and it would be beneficial if their training were directed towards this type of work. Table 8.1 gives a general picture of the

Table 8.1. Labour force composition

Trade	Small contractor (less than 50 men)	National contractor (maintenance dept)	Local authority (direct labour dept)
	%	%	%
Carpenter	25	26	16
Bricklayer	22	11	10
Plumber	4	9	10
Plasterer	1	2	6
Painter	26	39	34
Others	2	1	4
Handyman	1	1	2
Labourer	19	11	18
	100	100	100

composition of the labour forces of the three main types of organisation concerned with the execution of the maintenance work.

The higher proportion of bricklayers employed by small contracting firms is due to the fact that such firms undertake a certain amount of minor new works in the form of alterations and extensions and a lower proportion of painting and decorating than the other two types of organisation. It is usual for maintenance operatives to work in small gangs made up of a number of different trades. The BRE study of operatives' work in this field[3] revealed a good deal of overlapping between trades and also suggested that the relationship between a tradesman and the corresponding trade labourer was less clearly defined than in new work. The study revealed a large area of general maintenance and common work which defied normal trade classification and which was performed indiscriminately by most trades and their associated labourers. There were two exceptions to this flexible approach to trade demarcations:

(a) In Scotland and the northern regions operatives tended to restrict their work to that identified as the primary work of their trade.
(b) Local authority employees generally showed a tendency to specialise within their own trade.

However, it is clear that rigid trade demarcations can hinder the economic carrying out of maintenance work. To quote from the evidence submitted to the Phelps Brown Committee[4] by the Institute of Municipal Building Management, 'The encouragement of multi-craft skills could play an important part in controlling the rising cost of maintenance, by the introduction of a "building trade technician" recognised by the trade unions as an operative who in day-to-day maintenance is permitted to undertake work in more than one trade'.

It is generally recognised that maintenance requires greater skill than new construction work and the previously mentioned study of operatives' work demonstrated that in fact maintenance operatives were better qualified than those in new construction. In maintenance about 60 per cent of tradesmen under 30 years of age had a craft certificate and six out of ten of these had achieved advanced craft certificate level. The corresponding proportions in new construction were 38 per cent with a craft certificate, three out of ten having gained an advanced certificate.

Unfortunately, the wage structure laid down by the National Joint Council for the Building Industry is based on a simple division between craftsmen and labourers with only a limited number of plus rates in between for certain semi-skilled operatives such as steelbenders, scaffolders, plant operators, etc. In maintenance at least, it could be argued that there is a need to formulate a pay structure which will reflect the gradations of skill involved, so that the versatile craftsman can be more adequately rewarded. Generally plus rates, i.e. payments above the basic wage rate, are very much higher for new construction than for maintenance and are unrelated to skill and, in many cases, to productivity. There is little recognition of the fact that in maintenance there is an emphasis on an operative's responsibility and initiative which often requires him to work without much supervision and with a minimum of instruction and technical guidance.

Age and stability

The labour force in maintenance and repairs tends to be older and more stable than that in new construction works. The aforementioned study of operatives' work revealed that the proportion of men over 40 years of age was 58 per cent for maintenance as compared with 32 per cent for new construction. Since the proportion of apprentices is higher in maintenance than in new construction, there would seem to be a drift from maintenance to new construction shortly after the completion of training and a reverse movement after the age of 30. This is most pronounced for bricklayers and plumbers and to a lesser extent for carpenters and painters. The reason for the reverse movement of craftsmen is probably quite simply that as they become older and acquire family responsibilities, the greater stability, local nature and less rigorous working conditions in maintenance outweigh the higher bonus earnings obtainable in new construction. This seems to be borne out by the finding that only 25 per cent of operatives employed by small firms associated with maintenance were attracted to other jobs by incentive payments, as compared to 59 per cent for large firms mainly concerned with new construction. The emphasis placed on stability of employment is brought out even more clearly in the case of local authority direct labour forces which usually have a much lower labour turnover even than small building firms. Other factors which attract the older operative to maintenance may be the greater responsibility and sense of satisfaction in the work done than is usually possible in new construction, coupled with the greater variety of maintenance jobs and the personal contacts with building owners and tenants.

While the above would suggest that maintenance operatives are not so highly motivated by financial reward, it is more probable that the deciding factors are the disincentives of working on new construction sites. If these disincentives were removed by changes in employment policies to give greater continuity of employment and better working conditions, the drift to maintenance in later life might well be restricted. The balance could be preserved by making the rewards and conditions in the maintenance sector more attractive so as to produce a corresponding reduction in the movement of the younger craftsmen to new construction.

Maintenance technician

It is generally conceded that it would be desirable to have a specially trained maintenance operative. A practical difficulty in introducing a separate category of 'maintenance technician' is that it would cut across the traditional methods of grouping craft skills and would therefore probably meet with union opposition. The present groupings are:

1. By material. Traditionally trades are associated with a particular material, e.g. plumbers with lead, carpenters with wood, etc.
2. By tools or plant. This is really complementary to the first grouping in that special tools have been developed to shape and fix particular materials. However, dexterity in the use of a particular tool may enable a craftsman to undertake work outside his normal field, e.g. a trowel is used for both bricklaying and plastering and, therefore, a grouping of 'trowel trades' would cover both of these activities.
3. By function. There is a trend for trades to identify with the provision and maintenance of a functional element or system rather than the traditional material, e.g. plumbers with water services, although it is now usual for materials other than lead to be used.

4. By operational method. This grouping has emerged in response to the need for special skills for the erection of industrialised building systems. In a number of cases these methods are unique to the particular system and are not based on the traditional craft skills.

The above groupings are far from ideal from a maintenance point of view. For this purpose, it would be more useful to have a grouping based on the work to be carried out at a particular location, e.g. piecing in a new section to a defective window frame, including carpentry, painting, making good plaster reveals and perhaps reglazing.

This would raise the question as to whether a maintenance technician should be treated initially as such or should have additional training after acquiring skill in one of the main trades. One effect of creating a specialist maintenance technician would be to restrict the free movement of operatives between new construction and maintenance. This might be thought disadvantageous in that the movement from one construction activity to the other provides a ready means of adjusting to fluctuations in the demand for new construction. However, this situation accounts for the unfortunate 'casual' nature of employment on new construction sites and also results in neglect of the existing stock of buildings and higher maintenance costs when there is a heavy demand for new buildings. It is in fact a consequence of an imperfect system and could be avoided by rationalising the demand for new construction so that continuity of work is provided for a relatively stable labour force. Indeed, the greater industrialisation of building involving high capital investment in equipment will depend to a large extent for its success on an assured market. Also, industrialisation will lead to a divergence between the skills required for the initial fabrication and assembly and the subsequent maintenance. This is seen most clearly in the case of manufactured products such as motor cars and is likely to apply equally to factory-produced buildings.

The ensuing polarisation of the construction industry will emphasise the need for specially trained maintenance technicians and a separate wage structure commensurate with their skills and responsibilities. Additionally, the formation of single-purpose organisations concerned solely with maintenance would foster the development of more sophisticated planning techniques for the grouping of work so as to achieve economies of scale.

Factors affecting output

Before considering means to improve output it is necessary to examine the factors which account for the relatively low productivity in maintenance.

1. The small scale of the individual jobs, resulting in a high proportion of non-productive time. The time spent in collecting stores, preparing to do the work and clearing up on completion is substantially the same irrespective of the size of the job.
2. The diversity of the work content of the jobs. Even quite small jobs can involve a number of separate trades, leading to delays and additional cost if demarcation rules are strictly adhered to.
3. The non-repetitive nature of much of the work. Each job presents a fresh set of problems and may require novel and unfamiliar methods of working. A BRE study[5] of repetitive house construction showed clearly the increase in productivity during the learning period.
4. The adoption of inappropriate methods. This follows from the previous point in that the lack of continuity does not permit the development of optimum methods. The effect in this case is not limited to increasing the cost of the particular operation, but may also have an adverse effect on the long-term durability of the building and the cost of subsequent maintenance.
5. The dispersal of sites. The proportion of non-productive travelling time is related to the size of the job and the distance between site and depot. Also, the higher the dispersion factor the more difficult and expensive it is to provide effective supervision. An additional factor, particularly in relation to housing maintenance, is the time wasted on abortive visits through failure to gain access to the premises.

6. Difficult working conditions arising from:

   (*a*) Need to work in confined positions.
   (*b*) Inadequate means of access.
   (*c*) Restrictions on working times or methods.
   (*d*) Protective and safety measures necessary in an occupied building.
   (*e*) Inclement weather in the case of external works.

7. The making good of damage caused by the removal of old components and installations and the extra care necessary to limit consequential damage to a minimum.
8. Poor communications between site and office, particularly likely where sites are scattered over a wide area. The result is time wasted while operatives wait for instructions. A parallel consideration is that of communication between the client and the contractor to keep the client fully informed of progress.
9. Limitations on the use of mechanical plant imposed by the small scale of the operations.
10. The tendency to execute work only when it becomes a matter of urgency. This not only has a disruptive effect on the organisation of the labour force, but may result in deterioration of adjoining elements necessitating much higher expenditure than would have been necessary if the defect had been remedied earlier.
11. The difficulty of drawing up a detailed programme of work before a fairly extensive exploratory survey has been carried out to determine the extent of the damage, e.g. dry rot.
12. Obsolete or obsolescent materials which may be required to match existing work and which become increasingly expensive.
13. The poor management quality of many small firms engaged in this type of work and the lower calibre of professional support.
14. The age distribution of the operatives. It has been noted that the older craftsmen tend to gravitate to maintenance because they are attracted by the slower tempo of the work, while at the same time firms engaged in this work employ a higher proportion of apprentices.

15. Motivation. Many different theories of motivation have been proposed but none is entirely satisfactory. Hertzberg proposed that workers had two sets of needs, those which produced job satisfaction and those which produced dissatisfaction. The former which he described as motivators included achievement, recognition, the work itself, responsibility and advancement. The latter which he described as hygiene factors were largely to do with the working environment and included company policy, relations with supervisors and working conditions. He found that the factors which caused job satisfaction were quite separate and distinct from those which caused dissatisfaction. Thus, improving hygiene factors whilst removing a source of dissatisfaction did not necessarily act as a motivator. Maslow postulated a hierarchy of needs as illustrated in Fig. 8.1 and that needs only act as motivators when they are unsatisfied. Thus, the lower physiological needs are dominant until satisfied and then the next lower becomes dominant, and so on.

Fig. 8.1

Greater job satisfaction can be provided by job enlargement. This may take the form of horizontal job enlargement in which the worker is involved in a wider range of operations at the same level in the hierarchy or vertical job enlargement, also called job enrichment, in which certain aspects of the supervisory function are delegated to the worker. Participation may also be regarded as a form of job enlargement in that workers are involved in discussions on such matters as standards and methods of working. It is important in such situations that the individual is given sufficient information to allow him to participate effectively.

Incentive schemes

An obvious way to improve output would appear to be the provision of some sort of incentive to work harder. The objects of incentive schemes are stated in the National Working Rules for the Building Industry as:

1. To increase efficiency, thereby keeping costs at an economic level.
2. To encourage greater productivity, thereby providing an opportunity for increased earnings by increased effort while maintaining a high standard of workmanship and avoiding a waste of labour and materials.

However, one should examine the reasons for attempting to increase efficiency and particularly, as far as direct labour is concerned, whether the intention is:

1. to do more work with the same labour force, or
2. to do the same work with a reduced labour force, or
3. to deal more quickly with work orders.

In a reported case[6] a local authority achieved an increase in productivity which resulted in 300 doors a year being painted instead of, as formerly, 200. However, since the doors were already being painted sufficiently often, there was no real increase in efficiency.

It is also clear that in some cases increased wages are paid for the purpose of attracting and retaining operatives and any increase in output which results is quite fortuitous. Local authorities are at a particular disadvantage when competing with contractors for scarce labour in that, unlike contractors, they are unable to pay 'plus' rates. Any additional payments must, therefore, be made under the guise of an incentive scheme. From the contractor's point of view, the value of incentive schemes will be judged by the effect on profits. As many small maintenance jobs are let on a 'daywork' or 'cost-plus' basis (i.e. the contractor is paid the actual cost of labour and material plus a percentage to cover overheads and profit), there is apparently no gain from carrying out the work more economically. Indeed, it could be said that the more inefficiently the work is executed, the higher the cost and the greater the contractor's profit. In a survey carried out by the Advisory Service for the Building Industry[7] some firms admitted that they were not interested in improving productivity on daywork where their profit margins are assured. It was remarked that although it is against the interest of the general public, this mistaken attitude is probably widespread.

Incentive schemes may be broadly divided into those which provide a direct incentive by setting targets for the execution of separate parts of the work and those which provide an indirect incentive by attempting to involve operatives in company policy.

Direct incentive schemes

1. Piecework
The target is a lump sum payable for completion of a specified task irrespective of the time taken. It is thus possible for an operative to earn less than the standard basic wage and the scheme is usually confined to labour-only subcontracting in new construction work. The view has been expressed that cash targets do not have the same incentive effect as hourly targets and that they are probably more suitable for the smaller than the larger firms. While it is difficult to reconcile this statement with the rapid growth in recent years of labour-only subcontracting, it may be true in respect of the older operatives forming the bulk of the maintenance labour force.

2. Spot bonuses
These are offers of additional sums of money over and above the basic rate and are awarded mainly by working principals for urgent work. By their very nature they are not applied continuously and therefore have little effect on long-term productivity.

3. Targets related to estimate
Here the targets for the various operations are derived from the

rates used in building up the estimate. The major defect of this system is that where the estimates are competitive the rates will fluctuate according to market conditions and will not provide a uniform basis for output-related earnings. However, if the rates are adjusted the scheme can be comparatively cheap and simple to administer and can provide a continuous check on profitability and a good basis for labour control.

Usually a percentage is deducted from the estimated time to create a reserve during favourable times to counterbalance losses during unfavourable times. A problem associated with this method is that in practice the labour content of the rate is based on experience, perhaps aided by past records of the time taken to perform similar work, and in a number of cases it is little more than a guess. This is particularly so where the estimator is remote from the site and is unable to check the methods of working. In such cases the estimator proceeds by trial and error – if the workmen complain that they are unable to earn bonuses, he assumes the target is too 'tight', while if excessive bonuses are earned he assumes that the targets are too 'loose'.

4. Targets based on work study

Although work study is based on the parallel techniques of method study and work measurement, it is rare to find method study applied to maintenance. No doubt the main reasons are the non-repetitive nature of much of the work and the traditional nature of the craft skills employed. However, the first difficulty could be overcome to a certain extent by grouping small items of work and the second by a reappraisal of the skills required for maintenance work.

Having ascertained the facts, the basic questions to ask are:

What is done?   Is it necessary?
Where is it done?   Need it be done at that place?
When is it done?   Need it be done at that time?
Who does it?   Could it be done better by someone else?
How is it done?   Is there a simpler and more economical way?

Work measurement is very largely limited to large direct labour forces employed by local authorities. The building up of a library of standard times would be beyond the capacity of a small firm and would involve excessively high administrative costs. A national library of output values would be of some assistance, but differences from job to job and from region to region are such that the times would require adjustment for local conditions.

Work measurement consists essentially of breaking down a job into its constituent elements of work and repeatedly timing the execution of these elements. An element should be a discrete operation or task which can be easily identified and which has a definite beginning and end. As the element is being timed the operative's speed of working is assessed or 'rated' in relation to a predetermined standard known as 'standard performance', which is represented by 100 on the British Standard Rating Scale. This is defined as the performance which a worker of representative skill and experience can achieve without over-exertion as an average over the working day when motivated by an appropriate incentive. On this scale, 'normal performance' is represented by 75 and is the performance achieved in the absence of an incentive.

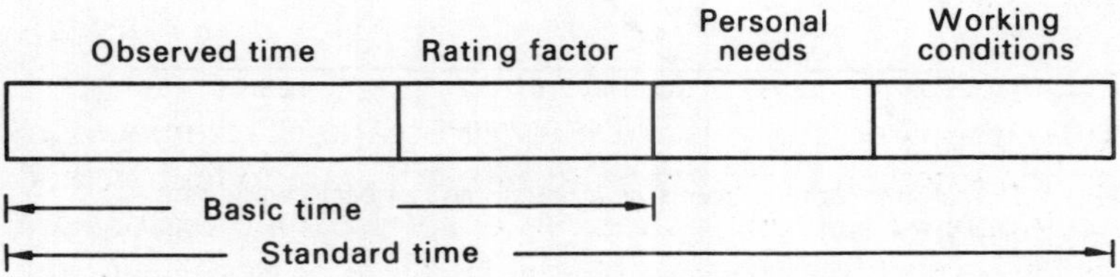

Observed time	Rating factor	Personal needs	Working conditions

|← Basic time →|

|← Standard time →|

Fig. 8.2. Calculation of the 'inclusive standard time'

The observed times are adjusted to give the corresponding times for a rate of working of 100 P and the result is called the 'basic time'.

$$\text{Basic time} = \text{Observed time} \times \frac{\text{Rated speed of working}}{100}$$

Finally, allowances for personal needs and for working conditions are added to the basic time to give the 'inclusive standard time'. These allowances are of necessity subjective and may form more than 50 per cent of the standard time (see Fig. 8.2). Provision must also be made for time spent in travelling to and from the job, and for preparatory work and clearing up afterwards. As this is related to the size and location of the job, it is better dealt with on a job basis rather than by inclusion in the elemental standard times.

Where the operative receives less than the full value of time saved on the target time, this is known as a 'geared incentive'. For most construction work a 50 per cent gearing is adopted, in which the operative receives half the value of the time saved. He starts earning bonus at 50 performance, reaches $33\frac{1}{3}$ per cent of his basic rate at 100 performance and continues to rise proportionately thereafter, as illustrated in Fig. 8.3. In the 50

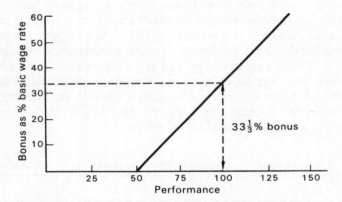

Fig. 8.3. Geared incentive: bonus as a percentage of basic wage rate
$= \frac{2}{3}(P - 50)$

per cent geared system the target is calculated by adding two-thirds to the standard hours. Thus, assuming a work content of 36 standard hours, the target would be $36 + 24 = 60$ hours. If the operative were to achieve a performance of 100 P and complete in 36 hours, the saving would be 24 hours, of which he would receive half the value (12 hours) representing a bonus of one-third of his basic pay.

In view of the uncertainties inherent in predicting target times for maintenance work, a gearing similar to that illustrated in Fig. 8.4 is sometimes adopted. Bonus payments start at 50 per cent standard performance and rise rapidly to 25 per cent basic pay at standard performance and then at a diminishing rate thereafter. This arrangement softens the effect of gross discrepancies in fixing targets and discourages workers from skimping work so as to earn excessively high bonuses.

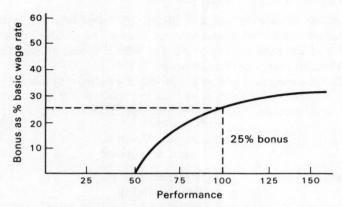

Fig. 8.4. Alternative geared system: bonus as a percentage of basic wage rate $= 50 - \dfrac{2500}{P}$

Indirect incentives

1. Profit-sharing
This may take the form of:

(a) Immediate profit-sharing, in which the gross profit is calculated on the completion of each job and a percentage of the profit is shared out among the operatives according to the time spent on the job.

(b) Delayed profit-sharing, in which the overall profits are calculated six-monthly, annually or over some longer period and a proportion is shared out according to some formula. Although immediate profit-sharing provides a greater incentive, delayed profit-sharing is probably more useful in retaining skilled operatives by involving them in the long-term profitability of the firm.

2. Merit rating
This usually amounts to a selective form of 'plus' rates in which additional payments above the basic rates are made for defined factors such as level of output, quality of workmanship, length of service, etc. The extra payment tends to become a permanent addition to the operative's wage and would be difficult to reduce if for any reason the operative's performance deteriorated. It is unlikely that merit rating has any great effect on

output but it does provide a means of rewarding the more competent and conscientious workmen, especially where employed on high-quality work for which speed of working would be an inappropriate measure.

The following is a possible method of grading payments:

Skill	Technical knowledge	10
	Versatility	10
	Quality of work	10
	Speed of working	10
Reliability	Timekeeping	10
	Responsibility	10
	Initiative	10
	Co-operation	10

Length of service

$$0-\tfrac{1}{2} \text{ year} \quad 5$$
$$\tfrac{1}{2}-1 \quad 10$$
$$1-5 \quad 15$$
$$\text{over } 5 \quad 20$$

20

──

100

Grade	I	0–25	Additional payments
	II	25–50	related to comparable rates
	III	50–75	paid by other firms in the
	IV	75–100	district.

The advantage over the blanket 'plus' rate is that additional payments are related to identifiable qualities, and, over the work-study-based scheme, that speed of working is not the sole criterion.

3. Percentage of turnover

This is confined to supervisors to encourage them to organise and control their work force to greater effect. It is an attempt to resolve the dilemma created by incentive schemes that if the supervisor is included in the scheme he has an incentive to pass bad work, while if he is excluded he may well be earning less than the men under his control.

4. 'Plus' rates

These are indiscriminate additional payments in excess of the basic rates laid down by the Working Rule Agreement, paid to all operatives merely for the purpose of recruiting and retaining an adequate labour force. They are not related in any way to output and clearly do not provide an incentive to improve productivity.

Effectiveness of work study

Work study has been singled out for more critical scrutiny in that the comparatively low level of productivity in building is frequently attributed to the absence of such techniques. However, it is misleading to suggest that the introduction of work study in itself would do anything to improve the situation. It is much more probable that schemes have failed through lack of attention to the wider aspects of planning and deploying the labour force to the best advantage. Incentive schemes based on work study may have certain failings in the context of maintenance and their effect should be carefully considered when introducing such a scheme. The main criticisms are the following:

1. There is frequently a confusion of aims between bargaining for a basic minimum wage and negotiating a proper reward for extra effort. This is likely to continue while basic wage rates are substantially lower than the actual wage that a man can command in the open market. Thus, the prime objective of many schemes is to attract and retain labour, and increased output merely covers the cost of operating the scheme. This was borne out by a study of local authority schemes[6] from which the following statements have been extracted:

   'This would mean that individual productivity has increased but that, because of higher wages, there is no saving to the Council.'

   'It would appear that the authority has not lost money by introducing the scheme since a small but definite improvement in output per man has just about balanced the extra costs involved.'

   'The total resources remain substantially unchanged, the reduction in site labour being balanced by an increase in administrative and supervisory labour.'

2. It is generally accepted that the keenest incentive is one based on individual effort, but it is often difficult to adopt such a scheme for building maintenance unless there is a high proportion of single-trade jobs. In addition to being administratively expensive, it could cause friction if there were wide disparities in the amount of bonus earned by the individual members of the small gangs employed on this type of work. It is much more important to encourage a spirit of co-operation even though this may result in some dulling of the incentive. However, the unit for bonus purposes should be as small as possible. Gang payments are particularly useful on programmed maintenance and where scattered groups of buildings have to be maintained. By this method each gang would pool their bonus which would be shared according to the number of hours worked.

3. There is little evidence that incentive schemes take into account the intensity of physical effort or of mental concentration. It is quite noticeable that some trades can earn more than others for the same effort, in some cases by ignoring safety measures which must be taken into account when fixing standard times. Also, only a small percentage of times are derived from actual site studies and where past records are used they rarely give sufficiently detailed information for the precise assessment of work content and working conditions. Indeed, it would be difficult to observe sufficient cycles of similar work to obtain completely reliable results and it is most unlikely that the savings would justify the exceedingly high cost involved.

4. Payments are usually made on the basis that the standard of work is satisfactory and that no remedial work will be required later. This appears to be an over-optimistic view, especially in those cases where the supervisor responsible for quality control is himself receiving a bonus based on the output of his men. The general principles laid down in the National Working Rules provide that where work proves to be defective the remedial work shall be carried out by the same gang, that no bonus shall be paid therefore and that the time taken shall be offset against any savings on subsequent targets. However, if, as often happens, the defect does not become apparent until some time after completion, members of the original gang may no longer be

in the employment of the firm. The greater the time lag between completion and discovery of the defect the more difficult it is to apply this principle. The main safeguard is efficient quality control of the work and owing to the dispersal of maintenance sites this is only possible by increasing the number of supervisors.

5. Work-study-based schemes do not take into account the level of responsibility which the operative may be capable of accepting and are based on the sole criterion of output. This limited approach to the work may not be in the best overall interests of the organisation and certainly does little to provide an outlet for creative self-development. It also ignores the fact that maintenance attracts the more mature craftsman who has a greater responsibility for the standard of his work.

6. Although schemes based on work measurement may tend to stimulate some degree of inventiveness on the part of the operative they also inhibit the disclosure of innovations due to the possible retiming of operations.

7. Such schemes do not reflect the overall objectives of the company and form no objective basis whatever for reconciling employer and operative interests. Indeed they tend to obscure the identity of interests to the extent that when an operative takes a day off work he sees the loss as his own and does not consider the disruptive effect on the work of the gang.

8. An incentive scheme is regarded by some firms as a substitute for management and too much reliance is placed on purely negative feedback from the site. Action is taken only in the event of feedback indicating that target times are too 'tight' or too 'loose'.

9. Many such schemes are too complicated for the operative readily to relate a given effort to the corresponding reward. If incentive schemes are based on the premise that a man will not work without a carrot, then much of the effectiveness is lost if the carrot is not clearly in view. In some cases work is post-targeted and cannot have any effect whatsoever on the performance of the operatives. In addition, the lack of interest engendered results in little effort being made to simplify methods or to avoid unnecessary non-productive time.

10. The assumption underlying incentive schemes is that all workers are highly financially motivated but it is doubtful whether this is true of maintenance workers. Clearly monetary reward is important, but there are other factors which may be of greater concern.

11. An operative may be penalised, i.e. deprived of the opportunity to earn bonus, by factors outside his control such as exceptional and unforeseen difficulties in carrying out the job, failure on the part of management to plan the work so as to avoid unnecessary delays in waiting for other trades, materials and instructions, and inclement weather. There is an understandable feeling of frustration and resentment among operatives when bonus earnings achieved during periods of high productivity are frittered away by stoppages arising from reasons beyond their control. The bad psychological effect which this induces can undermine discipline and lead to a general air of slackness which pervades the whole organisation.

12. The introduction of an incentive scheme based on work measurement calls for management skills which quite patently many of the smaller firms engaged on maintenance work do not possess. It is not surprising, therefore, that a study of 27 building firms[7] revealed that only one firm based targets on standard data. Admittedly, these were all small firms in which the labour force was possibly too small to bear the increased overheads which such a scheme would involve.

13. In the larger firms one often finds a division between the bonusing department and site management. It appears that bonusing is regarded as a completely separate activity which has no connection with planning and controlling the work. Unless the incentive scheme is an integral part of site management, its full potential as a means of improving productivity will not be realised.

14. A scheme will be successful only if it has the active support of the operatives. The major problems are:

    (*a*) Construction workers and maintenance workers in particular dislike being timed. In a reported case the resistance of the men was such that standard times were obtained from organisations in another locality where conditions might well have been different.

(*b*) Large fluctuations in earnings make family budgeting difficult and therefore schemes may be regarded with some suspicion, especially by the older operatives.

15. It is anomalous that overtime should be worked when a bonus scheme is in operation. If more than about 50 hours are worked in any one week the operative is physically incapable of performing at maximum efficiency with a resulting drop in bonus earnings.

16. Incentive schemes do not prevent the development of and may reinforce certain group 'norms' which militate against increased productivity. This is particularly important for such work as maintenance where the gang is usually small and fairly stable. A group consists of any two or more persons who have come together in a social framework and have agreed, usually tacitly, on certain ways of behaving, i.e. 'norms'. Such group 'norms' include standards of workmanship, speed of working, who is acceptable to the group, patterns of behaviour, etc. If they can agree only on one or two aspects, the group will be a 'loose' group, while if they agree on a wide range of behaviour patterns the group will be stronger and much more coherent.

Group 'norms' are important to people in that they give reassurance on concepts difficult to verify objectively, e.g. 'It is right to work hard all day'. A number of psychological experiments have been carried out to demonstrate the power of group 'norms' to influence individual action.[8] These indicate that the ways to prevent the development of 'norms' hostile to the smooth running of the firm are greater participation of operatives in decision-making and wider sharing of information. The work-measurement incentive scheme does not provide the opportunity for this type of involvement and can create an atmosphere in which operatives actively seek to exploit weaknesses in the scheme. The very small firm, perhaps with a working principal, is better able to encourage a spirit of co-operation as a result of the closer personal relationship between the proprietor and the men.

However, the foregoing is not to deny the usefulness of work measurement in the context of work planning and control and it could be said that incentive schemes based on work measurement are a necessary preliminary to the introduction of more sophisticated systems, such as productivity bargaining.

Productivity agreements

The aims of productivity agreements are defined in the National Working Rules as making a joint effort to improve efficiency by reducing unit costs through such means as the use of balanced gangs, greater flexibility or the relaxation of specified work practices. It is also emphasised that such an agreement should provide an opportunity for higher earnings.

On the other hand, the Building Employers Confederation (BEC) the National Federation of Building Trades Employers (formerly (NFBTE)), in its Points for Guidance of Contractors, stresses the importance of efficient site management to ensure that:

1. The quality and volume of output obtained in return for wages paid are satisfactory.
2. That all necessary information is available and planning is completed before an operation is started so as to avoid interruption of the work flow.

However, these statements do not convey the fundamental changes of attitude necessary for the success of a productivity agreement but tend to emphasise the dichotomy of interests – on the employees' side 'higher earnings' and on the employers' side 'volume of output'.

The implication is that increased earnings will produce worker satisfaction by giving him access to a higher standard of living. As opposed to this, it is assumed that by keeping wage rates to a minimum, costs will be reduced and the firm will be more competitive and profitable. However, the level of taxation on personal incomes, levies on the employment of labour, credit restrictions, the general economic position and opinions as to what constitutes an acceptable standard of living all serve to confuse the issues at stake.

In essence, a productivity agreement is a settlement in which the workers agree to make changes in working practices which will improve output and in return the employer agrees to reward the worker in some way, usually by increased wages. It follows, therefore, that the introduction of such an agreement will involve a careful assessment of the increased efficiency expected as a result of changes in the specified working practices so that the appropriate increase in wages can be calculated. It also calls for continuous control to ensure that the anticipated improvement in productivity is maintained.

Main advantages likely to flow from the adoption of a productivity agreement

1. A reappraisal of the structure of the organisation and the relations between management and men. In the long term this could be highly significant in leading to a situation in which no bargaining is necessary, but where standards of performance and co-operation are laid down against which a cash payment based on accessible company information is made. One would expect a natural consequence to be a more flexible classification of skills according to work content in place of the present rigid classification according to traditional crafts.
2. Better deployment of labour resulting from changes in customary work practices in relation to:

    (a) Demarcation agreements. It is probable that small building firms would gain little in this respect and that the relaxation of such agreements would be of principal benefit to local authority and other public direct labour forces. However, there would be more general benefits if this led to a more flexible approach which would permit each organisation to assess its own requirements in regard to the most suitable groupings of skills required in the particular circumstances.

    (b) Mates. The custom of certain craftsmen being accompanied by mates can be particularly wasteful in the case of small maintenance jobs and quite unjustified. Where part of the work requires two men, it can be arranged for other men to be working on the site at the same time to give necessary assistance, e.g. a joiner may help a plumber to lift a bath into position and the plumber can then make the connections alone. The attendance of one craft upon another in this way would make mates largely unnecessary.

(c) Non-productive time. It is a matter of observation that groups develop habits as to the length and frequency of breaks and the standard of timekeeping which seriously reduce the amount of time available for work. Other wasteful patterns might also have become normal with regard to the collection of materials, travelling from job to job and communication of instructions. An activity-sampling survey of a local authority maintenance department revealed the following pattern:

	Travelling			Waiting		
Average no. of men in sample	Depot to job %	Job to depot %	Job to job %	For materials or equipment %	For access to premises %	For instruction %
438	11·55	10·55	10·12	3·86	1·42	2·50

Thus only 60 per cent of the total time was spent on actual maintenance, the remaining 40 per cent being completely non-productive. Also some proportion of the 60 per cent would be devoted to breaks and personal needs so that probably less than half a workman's time was spent on actual work.

(d) Overtime. The object should be to reduce overtime by increasing productivity during the normal working day. There is a tendency for overtime to be regarded as some sort of reward with the result that work is carried out at a slower rate in the knowledge that extra hours at a premium will be available later. The objective thus becomes overtime and this is achieved by working at a slower pace during normal hours. Unfortunately, the slower pace is maintained during overtime working and it becomes a self-perpetuating process.

(e) Time flexibility. This is a parallel problem to overtime in that some maintenance departments may have to carry out certain operations outside normal working hours. This should not be regarded as overtime but catered for in the agreement by stipulating the number of hours per week and the proportion that the operative may be asked to work outside what is considered to be the normal working day.

plications and should not be considered in isolation. In particular, it demands a planned approach at all stages in order to provide the conditions under which the operatives have the opportunity to earn a reasonable bonus. There is little doubt that the most common factor which detracts from the success of incentive schemes is poor organisation resulting in discontinuities in the flow of work.

National Working Rules

The National Working Rules are published by the National Joint Council for the Building Industry which consists of representatives of the employers and operatives. The main functions of the Council are to fix the rates of wages of building trade operatives and to determine their conditions of employment. However, it is important to remember that within the general framework of the rules there are numerous regional or area variations and additions. The machinery of the Council includes conciliation panels at local, regional and national level to deal with disputes which arise under the Working Rules.

RULE 1 – Wages

This lays down that the standard basic rates of wages for craftsmen and labourers are to be fixed by the National Joint Council together with provision for a guaranteed minimum bonus payment. The concept of a guaranteed bonus was introduced in February 1973 to operate with effect from June 1973. Where an incentive scheme is in operation the provision applies only if the bonus earnings have fallen below the level of the guaranteed minimum for reasons beyond the control of the operative. Where there is no incentive scheme a sum equal to the guaranteed minimum bonus is to be paid provided the prescribed conditions of availability for work have been met.

The rule provides that persons being trained for a skilled building occupation should be trained under the terms and conditions laid down in the National Joint Scheme of Training and states the proportionate wage rates related to the age of apprentices and young male labourers.

Watchmen are to be paid at a rate per shift (day or night) calculated at one-fifth of the standard weekly rate for labourers. Trade charge hands and gangers are to be paid not less than the amount stated above the standard rate for craftsmen or labourers as the case may be. Qualified bar benders and fixers are to be paid the standard rate for craftsmen and scaffolders varying extra amounts above the labourer's rate according to whether they are trainees, basic scaffolders or advanced scaffolders. Both bar benders and scaffolders are required to provide proof that they have received proper training.

The rule also covers the scheme whereby the employer is responsible for purchasing holiday stamps of appropriate value from the Management company to cover the cost of annual holidays. Because of the casual nature of the employment it is probable that operatives will work for two or more employers during the year. To ensure that they get paid for their holidays in accordance with the Working Rules the operative is paid for his holiday by the employer for whom he is working at the time, the employer being reimbursed by the central body responsible for running the scheme. It should be noted that apprentices are not included in the scheme until their last year.

There is also a recommendation that an incentive scheme and/or productivity agreement should be introduced wherever practicable.

RULE 2 – Working hours

The rule provides that the working week shall be 39 hours – 8 hours per day from Monday to Thursday and 7 hours on Friday.

The rule also lays down that meal breaks shall not exceed one hour per day, at least half of which shall be the dinner interval. Starting time is a matter for local agreement but normal working hours shall not start before 07.30.

RULE 2A – Guaranteed weekly wages

Provided an operative is available for work throughout the normal working hours he is entitled to payment at his standard rate of wages for the full normal working hours whether work is or is not provided by the employer and regardless of temporary stoppages through inclement weather or other causes beyond the control of the parties. The provision does not apply if the employer is unable to provide continuity of work through industrial action. Also if the stoppage goes on for more than one week the employer may require the operative to register as an unemployed person.

RULE 2B – Conditions of service and termination of employment

This rule provides that an operative may be transferred from one job to another at any time at the discretion of the employer. The notice required to terminate the employment is related to the length of time that the operative has been employed by the firm.

RULE 3 – Extra payments

The types of work and conditions which entitle the operative to extra payment are divided into categories as follows:

A. Discomfort, inconvenience or risk
Work at heights
Furnace, firebrick and acid resisting brickwork
General (dirt, inconvenience or discomfort)

B. Continuous extra skill or responsibility
Timbermen
Mechanical plant operators

C. Intermittent responsibility
Scaffolder (part time)
Cleaning stonework

D. Tool allowances

RULE 4

This rule is divided into three sections.

A. Overtime – General provisions

Overtime shall not count until full time for the day has been worked and should normally not exceed one hour on Mondays to Fridays.

B. Overtime payments

For the five-day week overtime is paid at the following rates:

Mondays to Fridays – first three hours at time-and-a-half; afterwards until starting time next morning at double time.

Saturdays and Sundays – time worked from starting time on Saturday morning until 4 p.m. at time-and-a-half; from 4 p.m. on Saturday to normal starting time on Monday morning at double time.

C. Holidays

Work required to be done on:

(i) Christmas Day, Boxing Day, New Year's Day, the first Monday in May and any of the four days of public holiday fixed under the National Working Rules – days paid at double time.

(ii) Any other days which have been recognised by the Local or Regional Joint Committee as local holidays – days paid at time-and-a-half for the period of the normal working day.

RULE 5 – Shift work and night work

A. Double-day and three-shift working

This lays down the conditions which apply where either two or three consecutive shifts are worked each day. The normal hours of the shift shall be eight hours and the rates payable for the normal hours of the shift shall be plain time rates plus:

Double-day shift – a shift allowance of one-eighth of the plain time rate.

Three-shift working — a shift allowance of one-eighth for the first and second shift and one-fifth for the third shift.

Premium rates are payable where an operative is required to work beyond the normal shift hours and where shift working lasts more than three weeks the shifts should be rotated.

B. Night work

Night workers shall be paid an allowance of one-fifth of the plain time rate provided that at least three nights consecutively are worked. Normal working hours shall be eight hours per night Mondays to Fridays exclusive of meal breaks.

RULE 6 — Travelling and lodgings

A. Daily travelling

The terms 'district', 'free area' and 'place of original recruitment' are defined and the general provisions on daily travelling stated. Generally operatives set on at the job are not entitled to any payment in respect of daily travelling to and from the job. Where there is an entitlement to payment it is to be made at plain time rates unless the travelling time is greater than one hour, when the excess is paid at overtime rates.

B. Operatives sent out who do not travel daily

This rule applies where an operative is sent for a period to a job outside the district boundary and provides for the payment of fares to and from the job and at periodical leave intervals and for payment of travelling time and lodging allowances.

RULE 7 — Recognition of trade union officers

This rule covers recognition of both full-time trade union officials and site, job or shop stewards by the employer or his representative on site and sets out the procedures regarding meetings of workmen and inspection of cards. It also lays down notes for guidance in respect of safety representatives under the Health and Safety at Work Act.

RULE 8 – Payment for absence due to sickness or injury

This rule provides that where an operative produces satisfactory medical evidence and has satisfied the other conditions in this rule he shall be paid a stated sum for every working day that he is absent. For an operative who holds an annual holiday card current from the beginning of an accounting period the maximum number of days of absence shall be 48 days.

RULE 9 – Grievances, disputes or differences

This rule lays down the procedures to be followed for the raising and settlement of disputes.

General principles concerning incentive schemes and productivity agreements

The following recommendations are made by the National Joint Council for the Building Industry.

General principles
1. A target should be issued by management for each operation to be performed by an individual operative, or gang, and according to the extent that performance is better than the target, an additional payment should be made over and above the appropriate standard rate of wages.
2. Targets should be issued before operations are started and wherever it is possible to do so they should be agreed with the accredited representative of the operatives concerned, or with their union officer.
3. Targets should be based on standards of performance which have, wherever possible, been determined on jointly accepted work-study principles published by the BSI.
4. Targets are dependent on the saving rate adopted in each scheme. The incentive scheme must state the proportion of the saving which is to be paid out as bonus.
5. The number of operatives to be treated as a unit for bonus purposes should be as small as is operationally practicable. Bonus should not be paid on a trade or site collective basis

except where there are exceptional circumstances and it has been jointly agreed.

6. Incentive schemes should be expressed in simple and precise terms in order that (a) operatives may readily know what they have to do to increase their earnings and (b) misunderstandings and disputes may be avoided.

Operating principles

1. The target should be stated as a given quantity of work to be done in a given number of hours, to the satisfaction of management. (The given number of hours may be expressed as a monetary value where this method is customary.)

2. Where tasks are pre-measured they should be of short duration so that, as far as possible, they do not extend into a second payweek.

3. Gains and losses occurring in different payweeks shall not be offset, except where a target which has been pre-measured covers work to be done in more than one payweek.

4. Working targets once fixed may not be altered unless there is a significant change in the job content or in working methods and then only after joint consultation.

5. At the commencement of repetitive work a jointly agreed 'learning-curve' allowance is permissible having regard to the improvement in productivity that should subsequently follow.

6. The target will be inclusive for craftsmen and labourers and all hours will be chargeable against the target except where there is an interruption of work beyond the control of the parties.

7. The time of non-working supervision should not be charged against the gang. In the case of part-time working supervision the proportion of time to be charged against the gang should be agreed in advance.

8. The time of first-year apprentices should not be charged against the gang. In the case of apprentices in their later years of apprenticeship the proportion of their time which should be charged should, as a guide, be the same as the proportion of the craftsmen's rate which they receive under the apprentices' wage-for-age scale.

9. Overtime premiums, guaranteed time and travelling time should not be charged against targets.
10. Bonus payments, after adjustments in the case of a proportionate scheme, should be made at the standard plain time rate of the operative concerned, including extra payments under NWRs 3B and 3D.
11. The amount of bonus earnings should be notified to operatives not later than the payday immediately following the payweek in which the work was completed. The bonus should be paid not later than the next payday after that.
12. Where work for which bonus has been paid proves defective and has to be re-executed in whole or in part (i) the remedial work shall be carried out by the same operative gang, (ii) no bonus shall be paid therefor and (iii) the time taken shall be offset against any savings on subsequent targets. This provision shall not apply where the original work had been carried out strictly in accordance with precise instructions.

Safety and welfare

Construction regulations

The first code of regulations specifically dealing with building operations was made in 1926 but they were very limited in both scope and application and were eventually replaced by the Building (Safety, Health and Welfare) Regulations, 1948. The postwar expansion of the construction industry was coupled with an equally large increase in the accident rate and this, together with a need for improving safety in the civil engineering sector of the industry, called for more comprehensive legislation. This has resulted in the replacement of the 1948 Building Regulations by a group of four codes known as the Construction Regulations.

The Construction (General Provisions) Regulations, 1961 (S1 1961, No. 1580)
These regulations contain provisions dealing with rather specialised operations such as excavating, tunnelling, demolition, use of explosives, use of rail transport and others.

One very important requirement is that every employer of more than 20 men on construction work must specifically appoint a safety supervisor, or safety supervisors, to advise on the requirements for safety. This appointment is not necessarily a full-time one and whether it is or not will depend to a large extent on the size of the firm and the type of work carried out.

Some further requirements are that measures must be taken to prevent inhalation of dust, e.g. from cleaning processes and fumes (including exhaust gases from internal combustion engines), that all dangerous parts of machinery are properly guarded, that suitable lighting must be provided at all working places and at the approaches to the working places, and that steps must be taken to prevent danger from live electrical cables.

The Construction (Lifting Operations) Regulations, 1961 (S1 1961, No. 1581)

These regulations deal with the operation of different types of lifting equipment used for building work, including hoists, lifts, cranes and lifting tackle. In addition to controlling the use of such equipment they also require that it is kept in good order and safe for use.

The Construction (Working Places) Regulations, 1966 (S1 1966, No. 94)

These regulations are designed to ensure that a safe means of access is provided at every place at which a man has to work and that the place itself is made and kept safe. The regulations cover such items of building plant as scaffolding, cradles and ladders. They also regulate the methods of working in dangerous places such as on slippery roofs and on roof coverings of fragile material.

One of the most important requirements of the Construction (Working Places) Regulations is that no scaffold, including cradle scaffold, bosun's chairs and similar equipment, should be used unless it has been inspected by a competent person within the immediately preceding seven days. The object of such an inspection is twofold: firstly, to ensure that the scaffold is of adequate strength, sound material, properly installed and suitable for the purpose for which it is to be used, and secondly, to ensure that all safety devices required by the regulations, e.g. guard-rails and toeboards, are in position. Strict adherence to this regulation is the best way to ensure a reduction in this class of accident.

The Construction Regulations will be subject to amendment as necessary by new regulations made under the Health and Safety at Work Act.

Health and welfare conditions for the building industry

The Construction (Health and Welfare) Regulations 1966 now fall within the ambit of the Health and Safety at Work Act and deal with the facilities that should be provided for the health and welfare needs of employees on building sites. A simple interpretation of the regulations is incorporated in the National Working Rules covering the provision of the following facilities:

First-aid box
Where there are more than five men working on a site a first-aid box must be provided. If there are more than 50 men the man in charge of the box should have appropriate qualifications.

Shelters and accommodation for clothing and taking meals
On every site a contractor must provide shelter for protection against bad weather, personal and protective clothing and for taking meals. Where more than five men are employed the shelters must be warmed and have drying arrangements. Where there are more than ten men the mess room must have provision for heating food unless hot food is available elsewhere on the site. Drinking water must be provided at convenient points.

Washing facilities
Every site where anyone is employed for more than four hours must have washing facilities. Where there are more than 20 men or where the work will last more than six weeks the facilities must include (*a*) troughs, basins or buckets, (*b*) soap and towels, (*c*) hot and cold (or warm) water.

Sanitary conveniences
Every site must have one convenience for every 25 men up to 100 and one for every 35 men over 100.

Protective clothing
Protective clothing must be provided for men who have to continue working in rain, snow, sleet or hail.

Safe access to facilities
Proper paths must be provided to all shelters and conveniences and properly maintained.

Sharing of arrangements
A contractor may share facilities with another contractor on the site.

Contracts of employment

A contract of employment (or of service) is a contract between an employer and an employee (formerly referred to as master and servant) whereby the employee agrees to render some service in return for wages or similar remuneration. At common law an employer has a vicarious liability for the torts of his employees but not for those of independent contractors. It is important, therefore, to distinguish between persons who are employed under a contract of service and those who are engaged to provide services. The main test is the degree of control which the employer exercises over the manner in which the work is done. Thus, in general, the higher the degree of control the stronger the presumption that the person is an employee. However, the control factor may be outweighed by other factors, e.g. in the case of Morren *v.* Swinton and Pendlebury Borough Council 1965 it was held that in the case of a professional man the need to exercise detailed supervision was not so great as in the case of a manual worker and that as the council had appointed the plaintiff and had the right to dismiss him they were to be treated as his employers.

The Employment Protection (Consolidation) Act, 1978, brings together the statutory requirements which were previously contained in the Redundancy Payment Act, 1965, the Contracts of Employment Act, 1972, parts of the Trade Union and Labour Relations Act, 1974, and the Employment Protection Act, 1975. The Act requires an employer to give an employee written particulars of the main terms of his employment within 13 weeks of the commencement of the period of employment. This requirement may be satisfied by referring the employee to some other document which is reasonably accessible to him, e.g. the employer's handbook or a collective agreement. Unless the contract of service is for a fixed period there is a common-law right for either party to terminate the contract after due notice. The minimum periods of notice for employees who have been continuously employed for a period greater than four weeks is laid down in the Act as:

1. One week if the period of employment is less than two years.
2. One week for each year of continuous employment exceeding two years but less than twelve years.
3. Twelve weeks where the period of employment is twelve years or more.

Notice may be dispensed with where one of the parties has repudiated the contract, i.e. committed a fundamental breach which entitles the other party to treat the contract as at an end or if the contract is frustrated by some unanticipated event such as death of one of the parties.

An employee has remedies for dismissal both at common law and under the Act. At common law a person who is dismissed may bring an action for damages for wrongful dismissal. The statutory remedy is for what is called unfair dismissal and only applies where the employee has been employed continuously for 52 weeks for not less than 16 hours weekly or 8 hours weekly for a 5-year period. The employee must give notice of his complaint of unfair dismissal within 3 months of the dismissal taking effect. The onus is then on the employer to satisfy the tribunal that the dismissal was fair. This he may do by showing that the reason for dismissal was related to the capability or conduct of the employee or that the employee was redundant or that continuance of the employment would infringe some statute or that there was some other substantial reason. In the event of redundancy the Act provides that for every year of continuous employment (exceeding 2) between the ages of 18 and 21 the redundant employee is entitled to half a week's pay, for every year between 22 and 40 to 1 week's pay and for every year between 41 and 65 (60 for women) to 1½ week's pay. Wages over £152 per week and employment exceeding 20 years are ignored. An employee will not be entitled to the statutory payment if he is offered re-employment on the same terms, or suitable alternative employment if in either case he unreasonably refuses the offer.

Certain types of discriminatory treatment in the appointment of employees are now barred by statute. The Sex Discrimination Act, 1975, prohibits unfair treatment on the grounds of sex and marital status, the Equal Pay Act, 1970, guarantees equality of terms and conditions of employment for men and women and the Race Relations Act, 1976, prohibits discrimination on the grounds of race, colour and nationality. Complaints under these Acts are heard by industrial tribunals with provision for conciliation under the auspices of the Advisory, Conciliation and Arbitration Service (ACAS).

The Industrial Tribunals (Rules & Procedures) Regulations 1980 lay down new procedural rules which should reduce the time and expense involved in defending frivolous claims. The Employment Act 1980 provides for payment out of public funds in respect of trade union secret ballots and for the issue by the Secretary of State of Codes of Practice on such matters as closed shops and picketing. In the case of unfair dismissal the Act provides some relief for small firms by extending the qualifying period to 2 years provided that at no time during that period did the number employed exceed twenty.

References

1 Holmes, R. 'Feedback on housing maintenance.' BMCIS Occasional Paper, July 1983.
2 Ministry of Public Building and Works (1968) 'The use of information in the construction industry.' R. & D. Paper.

3 Building Research Establishment (1966) *Building Operatives' Work*. Vols. I & II. HMSO.

4 *Report of the Committee of Inquiry under Professor Phelps Brown into Certain Matters Concerning Labour in Building and Civil Engineering*. Cmd. 3714, HMSO, 1968.

5 Bishop, D. 'Labour requirements for house building – advantages of continuity.' Building Research Establishment Current Papers. Construction Series 18.

6 Local Government Operational Research Unit (1969) 'An introduction to incentive schemes in building maintenance.' Department of the Environment. R. & D. Paper.

7 Advisory Service for the Building Industry (1969) 'Incentive schemes applied to building maintenance of small firms.' Department of the Environment. R. & D. Paper.

8 Golembienski, R. T. (1962) *The Small Group: an Analysis of Research Concepts and Operations*. University of Chicago Press.

CHAPTER NINE

Contract work

Formation of contract

A contract for building or maintenance work is an agreement between two parties – the builder or contractor and the building owner or employer. The basis of the contract is the offer by the contractor to carry out the work for a certain sum of money and the acceptance of that offer by the employer. It is pertinent to note that the interests of the employer and the contractor differ and are in a sense incompatible. The contractor quite naturally seeks to obtain the contract at the highest price so as to maximise his profits, while the employer is anxious to keep his costs to a minimum consistent with an appropriate standard of workmanship. Clearly there is no advantage to the employer in securing a low price if the result is poor quality work. The essence of the agreement should be the payment of a reasonable sum for work which is properly executed and completed on time.

The offer is usually the contractor's tender and in the case of Crowshaw *v.* Pritchard 1899 it was held to be equally binding if described as an 'estimate'. The acceptance by the employer must be made within a reasonable time and must be unconditional. If it suggests new terms it does not create a binding contract. Although there is no legal necessity for building contracts to be in writing, for purposes of record and to avoid disputes it is highly desirable that they should be so evidenced.

It sometimes happens that after a contract has been formed

one of the parties discovers that it does not correctly represent his intentions. The most usual type of mistake is the discovery by the contractor of an arithmetical error in his tender build-up resulting in a disadvantageous price. In general, unilateral mistakes of this type do not entitle the party adversely affected to rectification. But where the parties are under a common misapprehension it is open to them to correct their mistake by mutual agreement. In construing a contract the court attempts to ascertain the meaning of the actual words used in the contract documents. Other documents which do not form part of the contract are inadmissible except in so far as they explain the circumstances surrounding the contract or throw light on any special meaning attached to the words used in the contract documents. Thus where quantities are supplied to assist the contractor in arriving at his tender price but do not subsequently form part of the contract, the builder will usually be unable to recover for indispensably necessary work undermeasured in the quantities. However, if the quantities are incorporated in the contract they become part of the description of the work and the value of any work omitted from or understated in the quantities is recoverable.

In the vast majority of cases contracts are discharged by due performance. However, a breach of a fundamental term by one party gives the other party the right to treat the breach as a repudiation and sue for damages. For a breach to constitute repudiation it must go to the root of the contract and this would include such actions as complete abandonment of the whole works before substantial completion by the contractor or failure by the employer to give the contractor possession of the site. In normal circumstances breaches such as delayed completion do not amount to repudiation although they may entitle the employer to damages, either liquidated damages agreed at the time of entering into the contract or unliquidated damages awarded by the court. The measure of damages would be losses which arise naturally from the breach or those of which the contractor had actual or imputed knowledge. Where the work has been completed substantially in accordance with the contract requirements the damages would be the cost of rectification or the difference between the value of the work done and the value of the work which should have been done. Thus in Hoenig v. Isaacs 1952, in a contract for the decoration of a flat and provision of certain fittings, the employer withheld payment on the ground that parts of the work were defective. The court held that there had been substantial compliance with the contract and that the builder was entitled to the contract sum less the cost of remedying the defects.

Other reasons which a contractor can advance for failure to complete the work include:

1. Impossibility. If circumstances become so radically changed that without default of either party the contractual obligations become incapable of being performed for reasons which could not have been foreseen by the parties at the time of entering into the contract both parties are automatically discharged. This is a rare occurrence on building contracts and would not normally cover such events as bad weather, inability to obtain adequate supplies of materials or strikes.
2. Illegality. If at the time of entering into a contract the performance is illegal or contrary to sound morals the contract is void. If its completion becomes illegal after execution of the contract the contract is frustrated and both parties are discharged from further performance.
3. Default of other party. As stated earlier if either party so acts or so expresses himself as to show that he does not mean to accept the obligations of a contract any further the other party may treat it as a repudiation of the contract and sue for damages. Examples of such acts are:

   (a) On the part of the employer – failure to give possession of the site, failure to make proper payments and wrongful dismissal of the contractor.
   (b) On the part of the contractor – complete abandonment of the work before substantial completion.
4. Waiver. This occurs where the parties either expressly or impliedly make a new agreement in place of the original contract, either before the work commences or during its progress.

In addition to the contractual liabilities it is also necessary to consider possible liabilities in tort. A tort is a civil wrong which occurs by reason of a breach of a general duty which is owed to society as a whole. It is therefore much broader than a contractual obligation which is owed only to the other party to the contract. The more common heads of liability in tort include negligence, nuisance and trespass.

1. Negligence. A professional man must apply a fair, reasonable and competent degree of skill and care in the exercise of his duties and the usual way of testing this is whether or not other persons of the same profession would or would not have done the same thing in the circumstances. This duty is owed to everyone who might be affected by the wrongful act. Recent judicial decisions have resulted in an extension of the time during which persons concerned with the design, construction or inspection of buildings may be liable. The Limitations Act, 1980, provides for a period of limitation of six years for simple contracts and twelve years for contracts by deed from the date of the cause of action. The date of the cause of action was originally taken as the date of completion of the contract but in the case of Pirelli General Cable Works Ltd *v.* Oscar Faber and Partners 1983, it was held that a cause of action in tort for negligence in the design or workmanship of a building accrued at the date when physical damage occurred to the building whether or not that damage could have been discovered at that date by the plaintiff. The uncertainty as to the total period of liability created by such judgements has been resolved by the Latent Damage Act 1986 which provides that legal action will be allowed over defects up to three years after they are discovered, or should have been discovered, with a 'long stop' rule which will bar actions more than fifteen years after the alleged bad work.

2. Nuisance. Building operations are often noisy and dirty and likely to give offence to adjoining owners. However, provided all reasonable steps are taken to minimise the inconvenience the neighbours must put up with it.

3. Trespass. Except for the many statutory organisations which have powers of entry no one is entitled without leave to cross the boundaries of another person's land. This includes placing or allowing materials to get onto another's land and the remedy for so doing is damages and/or an injunction to secure discontinuance of the act.

Generally one is not responsible for the torts of an

independent contractor but there are exceptions to this rule. Where the work is by its nature dangerous and likely to cause damage to other property the employer has a strict liability in the event of damage being caused and he cannot escape this liability by employing someone else to do the work. This applies particularly to work to party walls or where hazardous techniques are being used. In such cases it is important to ensure that the contractor has taken out insurance in the joint names of both the employer and himself.

Types of contract

The main differences between the various types of contract lie in the methods of evaluating the work and the degree of financial risk borne by the contractor and the client respectively.

Lump sum contracts

The contractor agrees to execute the work for an agreed sum based on information derived from drawings, specifications, bills of quantities or site inspection. The method presupposes that there is sufficient information on which to assess accurately the scope of the work involved. Small areas of uncertainty may be left to the discretion of the contractor to assess the risk and price accordingly or may be covered by the inclusion of a provisional sum. In the latter case, the cost of the work would be ascertained after execution on the basis of daywork, or in accordance with a schedule of rates submitted with the tender or by agreeing a reasonable sum. However, while isolated provisional sums are admissible for minor items, a disproportionate number of such sums would completely destroy the primary advantage of this type of contract, i.e. knowledge by the client of his financial commitment before the work is commenced.

Schedule contracts

This type is also called a measure and value contract and is useful where details are too scanty to permit the preparation of a precise specification at the time of commencing the work. The schedule lists all the items of labour and material which are expected to be required and may be an *ad hoc* schedule for a particular job based on past experience of similar jobs or a standard schedule designed to cover a wide range of jobs. The schedule may be unpriced, in which case the contractors tender by inserting a rate against each item, or there may be standard rates included in the schedule and the contractors tender by quoting a percentage on or off the standard rates. The work is measured on completion in terms of the schedule items and priced out either at the contractor's rates or at the standard rates plus or minus the quoted percentage. The cost is, therefore, not known until the work has been completed although the approximate quantities of the various items may be estimated prior to commencement for cost control purposes.

The PSA Schedule [1] is an example of a standard prepriced schedule and it has been suggested that it could be used with advantage by private firms. Although a convenient way of providing in advance for the pricing of work of uncertain scope it lacks estimating accuracy. The contractors tendering are required to state their offers in the form of one or two percentages on or off the schedule rates as a whole. The presumption is that an overall percentage adjustment will bring the schedule rates into line with those normally charged by the contractors. This is unlikely to be the case for the following reasons:

1. There is little uniformity among contractors regarding the pricing of individual items of work and inevitably some of the schedule rates will be higher and some lower than a particular contractor would normally charge.
2. The mix of items for a particular job may well exaggerate the effect of differences in pricing patterns. For example, a job may consist largely of those items for which the standard rates are higher than the contractor's usual rates and the addition of a percentage would serve only to make the discrepancy even greater.
3. The final cost will depend upon both the quantity and the rate for each item comprised in the work executed. If one accepts that some of the schedule rates will be 'high' and some 'low' then the relative quantities of each will affect the outcome.

4. The schedule rates are essentially averages and do not reflect the particular conditions under which individual items of work will be executed.
5. Standard rates tend to be updated at infrequent intervals and the pattern of pricing becomes increasingly inappropriate, due to differential changes in the prices of materials and new working methods.

In order to arrive at a realistic estimate, the contractor tendering should therefore estimate the likely proportions of the schedule items in the work which is likely to be ordered and then determine a percentage which will equate the cost based on the schedule rates to be cost obtained using his normal rates. Clearly this process is shrouded with uncertainty and it is probable that a contractor would merely identify the predominant trades and base his percentage on the estimated quantities of the major items in these trades.

However, in spite of the above defects, a schedule fulfils a number of useful functions. Principally it establishes standard work descriptions and lays down a pattern of pricing which, although not corresponding precisely to the pattern of prices normally adopted by the individual contractors, is sufficiently close to form an acceptable basis for valuing work executed. The percentage additions also provide a ready means of establishing trends in maintenance prices.

Cost reimbursement contracts

The contractor carries out the work and is paid the prime cost of labour and material plus either a previously agreed percentage or a fixed fee to cover overhead charges and profit. The obvious disadvantage of this type of contract is the absence of any financial incentive to encourage the contractor to carry out the work as economically as possible. Its use should therefore be restricted to small or extremely urgent jobs for which no other method is possible.

However, there may be occasions on which the nature of the work is so uncertain that, in an effort to provide cover against every eventuality, a firm price might be in excess of that obtained on a cost-plus basis. Also, although incentive may be lacking on an individual job the position would be different if there were the prospect of a series of jobs. Certainly this type of contract demands the employment of a reputable contractor and close scrutiny of the contractor's account to ensure that the labour and materials charged are reasonable in relation to the work done.

The additional cost of supervision is another factor that should not be overlooked. For lump sum contracts the supervisory function relates mainly to quality and checking that the work is being done according to instructions. For cost reimbursement contracts the supervisor's duties would extend to making sure that the work is being done in the most economical manner with the minimum wastage of materials. Such interference in the manner of executing the work is justified by the fact that extra costs arising from inappropriate working methods are borne by the client and not by the contractor.

Also, the practice of making the selection solely on the basis of the percentage addition quoted for overheads and profit is questionable. Clearly a contractor who quotes a marginally higher percentage than his nearest competitor might organise the work more efficiently and produce a lower prime cost which, in spite of the higher percentage addition, would result in a lower total cost. Indeed, it is probable that the higher percentage reflects a more sophisticated management structure which would be capable of achieving this desirable end. It is necessary, therefore, to ensure that there is equivalence of management skills before accepting the contractor who quotes the lowest percentage.

The main weakness is the difficulty of checking the correctness of the contractor's account, particularly with regard to the number of labour hours. A method that has proved beneficial is known as 'controlled daywork' whereby all jobs are pre-estimated prior to issuing the order. Then if the contractor's account exceeds the estimate by more than a certain percentage the reasons are investigated. The permitted variation may range from 10 per cent for the larger jobs to 20 per cent for the smaller jobs. If the estimates are based on the recorded costs for similar jobs with suitable adjustments for differences in working conditions they give a positive check on the reasonableness of the contractor's charges.

Further control is possible by more clearly defining prime cost following the lines of the RICS/BEC publication entitled

Definition of Prime Cost of Building Works of a Jobbing or Maintenance Character. This will remove any doubt as to what matters are included in the percentage addition and what can be charged for separately.

However, the fact remains that the major risks are borne by the employer, e.g. the extra cost of overcoming difficulties which impede the progress of the work are recoverable whereas in many cases these can be regarded as legitimate business risks.

Fixed price maintenance contracts

This method was developed by the then Ministry of Public Building and Works, Southern Region, in an attempt to reduce the considerable amount of paperwork involved in administering the other types of contract. In a number of cases the administrative costs exceeded the cost of actually executing the work and involved some 76 to 82 separate processes according to whether the work was carried out on a daywork or schedule basis.

The new system involves agreeing a lump sum with a contractor for undertaking a range of recurring works of a similar kind to a specified group of buildings over an agreed period. The method is somewhat similar to insurance in that the contractor agrees to carry out all the work of the specified types with no adjustment of the contract sum. Prior to the introduction of the system an analysis was made of the work carried out in maintaining a selected group of similar buildings and its cost over the previous year. Contractors then competed by quoting a percentage on or off the sum so ascertained. The contractor is paid one-twelfth of the annual lump sum each month although inspections are made from time to time to ensure that the work is being completed satisfactorily.

The advantages claimed for this system are:

1. There is a considerable saving in administrative time with the result that qualified personnel are released for other duties.
2. Payment to contractors is made promptly each month without retention and without the need for detailed measurements or checking accounts.

3. It is thought that there is some improvement in the quality of the work since it is in the contractor's interest to reduce the probability of expensive failures by early preventive maintenance.
4. It improves tenant satisfaction in that repairs are executed promptly and in some cases on the initiative of the contractor.
5. There is a direct incentive for the contractor to do the work as economically as possible.

However, it is probable that unless the number of units to be maintained is very large there will be significant deviations from year to year in the total amount of work required. There may also be differences of opinion as to whether a particular item of work is necessary. This could happen in those cases where a component deteriorates at a slow rate and has reached a point where, while still functionally satisfactory, it is considered by the user to be below an acceptable standard. Such disputes are likely to become more prevalent towards the end of the specified period.

Clearly the successful operation of such a contract calls for the utmost goodwill and understanding on both sides. The obligations of the contractor are open-ended in that in return for a fixed annual sum he agrees to carry out either on notice from the superintending officer or on his own initiative all work of very broad and ill-defined types, i.e. all cleaning of gutters and gullies, repair, renewal and maintenance work to the fabric, fittings, finishings, hot and cold water and heating services, and drains. There are certain exceptions, principally decorations where not associated with an item of repair or renewal and any single item exceeding an agreed value. Although an interesting development, it would seem that this type of contract is only suitable for small, repetitive items of work for which the need to repair is obvious and not dependent upon subjective judgment.

Term contracts

Under this type of contract the contractor is given the opportunity to carry out all work of a certain type or falling within certain limits of cost for an agreed period. The work done is usually priced on either a schedule (measured term) or a cost-reimbursement (daywork term) basis, although for the larger jobs it may be more advantageous to negotiate a lump sum.

The main advantages claimed for term contracts are:

1. Lower prices quoted by contractors in consideration of the benefits of an assured programme of work. This is probably true, but the estimating errors are likely to exceed any marginal allowance that might be made on these grounds. Where a schedule is to be used for the valuation of work executed, the quoted percentage addition must allow for fluctuations in wage rates and materials prices over the contract period (usually two years). As wage rates and individual material prices will vary at different rates and their proportions in the ultimate workload are not known, the probability that the percentage represents the correct amount is clearly very low.
2. Saving in time and overheads compared with that entailed in arranging single-job contracts. However, although there would obviously be some reduction in pre-contract time, the time required for the measurement of completed work for a schedule contract or for supervision of and checking accounts for cost-reimbursement contracts would be little affected. Where the work can be specified with reasonable precision a negotiated lump sum would probably give the greatest total saving in time.
3. The long-term relationship results in the contractor becoming increasingly familiar with the building and the needs of the occupants. In this it confers some of the benefits of a directly employed labour force, particularly if the same operatives are employed throughout the term.

The primary disadvantages are:

1. The reasonableness of the contractor's rates are tested in open competition only at infrequent intervals and may become increasingly unfavourable.
2. If a contractor's employment is terminated for any reason, a number of partly finished jobs will be left which will have to be completed by another contractor at an increased cost.
3. Towards the end of the contract period the number of

orders, especially for the larger jobs, will tail off and at the beginning of the next period will reach their normal level only when the new contractor has settled in.

However, on balance it would seem that the term contract has considerable advantages in the context of a planned maintenance system. The special relationship subsisting over a period of years permits the contractor to participate in the planning process and possibly suggest alternative timings for the work, resulting in some cost savings.

Usually contractors who have been previously vetted for inclusion in an Approved List would be invited to tender. The essential information that should be given to contractors tendering for term contracts is:

1. Form of tender and instructions relating to date and mode of submission and form of acceptance.
2. General Conditions of Contract to be used. Where these have been specially drawn up a copy should be supplied to the tenderers.
3. The contract area and details of the buildings included in the contract.
4. The contract period and commencement date. Usually the term would be from two to three years with provision for the annual updating of the rates on an agreed basis, e.g. by reference to a published index.
5. The likely total annual value of work that will be ordered. Although this is not legally binding it should be borne in mind that the contractor will have taken this into account when fixing his percentage addition to the schedule rates.
6. The maximum value of any single order above which a lump sum quotation may be required.
7. The method of valuation. The primary means of valuation will be by reference to the Schedule of Rates on which the contractor will be required to quote a percentage adjustment. The aim should be to cover 80 to 90 per cent of the jobs with the scheduled rates, with the daywork element restricted to those jobs which are of such an uncertain nature or executed under such diverse conditions that a schedule rate would be inappropriate.
8. Specification of materials and workmanship. This would follow the usual pattern of good building specifications but

covering items of unusual character particular to this type of contract, e.g. constraints imposed by working in occupied buildings.
9. Schedule of Rates. This may be a standard pre-priced schedule or one devised to meet the particular requirements of the building owner. In the latter case a schedule based largely on numbered composite items can reduce the administrative costs of measurement and valuation.

Suitability of contractor

The basic problem of selection is to predict which contractor of those available is most likely to achieve the client's objectives. It is necessary, therefore, to compare the contractor's known or assumed abilities with the services required of him. This process involves competition between those contractors who wish to secure the contract. The competition may be formal and based on criteria which can be objectively measured, or it may be informal and based on subjective judgment. Even where a building owner approaches a single contractor there will be some degree of implied competition in that the mere choosing of a particular contractor suggests that there are reasons for rejecting all the other contractors who are equally capable of undertaking the work.

The contractor's suitability should be considered under three heads:

1. Has he the potential resources necessary for the performance of the services required?
2. Is he likely to apply these resources adequately to the contract and are there any reasons why he should not do so?
3. What are the contractor's specific proposals for the contract and are they reasonable?

To determine the degree to which a contractor meets the above criteria, it is necessary to consider the following aspects of his organisation.

Reputation

The contractor should have a proved record of successfully performing the type of work required. A survey by the Research Institute for Consumer Affairs[2] revealed that in fact building firms were mostly selected on the basis of either personal knowledge or recommendation and that, by and large, clients were satisfied with the work done.

Financial stability

The ease with which anyone without capital or experience can set up in the building business has been commented on in many official reports. Equally, the very high bankruptcy rate is a matter of public record. It is quite clear that many small firms concerned with maintenance have very little working capital and that bad debts or long delayed payment can affect critically their financial situation. Most firms in this category are private concerns which rely on short-term loans and overdrafts from banks and on their own personal savings for finance. Further assistance is obtained from credit facilities provided by builders' merchants, and their willingness to supply materials is a fairly reliable indicator of the financial standing of the builder. However, in the absence of more detailed information, one would have to form an opinion of the builder's stability on the basis of such factors as length of time in business.

Resources

The contractor's resources may be considered under the following heads:

1. Physical – the nature of the premises from which the contractor conducts his business, including offices, workshops and stores. Also, the plant and equipment owned by the contractor and his access to building materials.
2. Human – the operative skills available in the contractor's organisation and their suitability for the work to be undertaken.

3. Management – the number and qualifications of managerial and supervisory staff and the extent of their technical knowledge and experience. This would be particularly important where the contractor is engaged for a term of years and is required to participate in the planning of the work.

Scope of work

This would involve an assessment of the type and size of jobs normally undertaken by the contractor and the work which he usually sublets. Firms can be broadly divided into general builders capable of dealing with all the traditional trades and specialist firms. General builders may be further subdivided into small firms capable of undertaking jobs up to about £500 in value and medium-sized firms for the larger jobs.

There has been a tendency in recent years for firms to specialise in certain aspects of what were hitherto regarded as general building, e.g. drain rodding, remedial treatment of timber, etc. By their greater experience in a limited field of operations they are usually able to offer a better service than a general builder.

Availability

There is little advantage to the building owner if the contractor possesses all the necessary resources but is unable because of concurrent commitments to use them when required. It not infrequently happens that small builders accept more work than they can conveniently deal with, to provide a backlog of several weeks' duration and so ensure continuity. This delayed response provides a useful means of smoothing the flow of work to the operatives and in an uncertain market enables the contractor to plan at least in the short term. For certain types of work the response time will vary according to the time of the year and may be very much shorter in winter. At such times the fall in demand which leads to a reduction in the response time also results in lower prices. There is thus a price advantage to be obtained from planning work so that it can be executed during a slack period, provided that it is not detrimental to the use of the building or to the quality of the work done.

Co-operativeness

Although a very difficult quality to assess in the absence of personal experience, it is essential that there should be harmonious relations between the building owner and the contractor. Often the nature of the work is not susceptible to precise definition at the time of selecting the contractor and the actual extent of the work may not become apparent until surface finishings have been removed. There is thus a need for close liaison between the client organisation and the contractor in devising methods which will be both technically and financially satisfactory.

Price levels

In many cases price is the sole criterion by which contractors are judged, but it should be noted that if the price is too low it may inhibit the attainment of the building owner's other objectives relating to quality and time. Orchard[3] makes the point that unless care is taken over the selection of the contractor, what appears to be a keen price can lead to a lot of subsequent trouble and expense.

The price is made up of the basic cost to the contractor of labour, materials and plant plus an allowance for profit. The basic cost is a reflection of the ability of the contractor to organise the work so as to reduce ineffective time to a minimum, while the profit element will to a large extent indicate his keenness to secure the contract and his workload at the time of tendering. In addition, the price will be affected by the skill of the contractor in estimating the basic cost and the price levels of his competitors.

Mode of selection

It is perhaps axiomatic to state that selection should be by those criteria which are least costly to apply.

Degree of competition

Tendering procedures may be classified according to the degree of competition as follows:

1. Open tendering. This is where a job is advertised and all contractors are free to quote without any prior enquiry as to their competence. The method has been widely used by local authorities in that it fulfils the requirement of public accountability and removes any suspicion of favouritism. However, the method is subject to criticism in that the low prices resulting from indiscriminate tendering are reflected in the quality of the work done and that resources are wasted when too many firms tender for the same job.
2. Selective tendering. In this method the client invites tenders from a limited number of reputable firms. The list of contractors may be drawn up specifically for a particular contract or the client may have a standing list of approved contractors from which a short list is drawn up for each contract. Separate lists may be kept according to the type of work or size of jobs which the firms normally undertake. Where a standing list is used it should be reviewed at intervals so that progressive new firms are not excluded and also so that unsatisfactory firms can be removed from the list.
3. Negotiation. This is where the client invites a tender from a single contractor who is known to have the necessary qualities, abilities and resources to carry out the work satisfactorily. Although there is no competitive element present and as a result the prices may be somewhat higher than with the other methods of tendering it permits a greater involvement of the contractor during the planning process and perhaps speedier completion and better quality work.

The number of stages

Broadly, methods may be single-stage, in which a single contractor is approached, or multi-stage, in which the characteristics of a number of contractors are compared in turn with criteria representing the building owner's requirements until eventually only one contractor remains as satisfying all the criteria.

The type of criteria used

This will vary from the single criterion of price to a combination of subjective assessments of the various characteristics of the contractors under consideration. The characteristics may be assessed on a numerical scale and each weighted according to its importance to the building owner. The weighted individual factors may then be totalled to give a single unit for the purpose of comparing the overall suitability of the different contractors.

The stage at which the contractor's offer is made

This will depend upon whether or not it is possible to specify fully the work before inviting tenders and starting on site. Where this is possible and fixed-price tenders are invited, the selection of the contractor and the acceptance of the offer occur simultaneously when a particular tender is accepted. Where the degree of uncertainty is such as to preclude this or where it is desirable that the contractor should give preliminary advice on the scope of the work, the selection of the contractor and the submission of an offer by that contractor will be separated in time. In this latter case, if price is to be a criterion it will be necessary when making the preliminary selection of a contractor to examine the methods which will subsequently be used for building up prices. Thus in addition to general information on the contractor's resources and reputation more specific information is required on percentage additions for overheads and profit.

The nature of the offer

This will be determined by the type of contract which the building owner has elected to adopt. The offer may be for a fixed price, either stated as a lump sum at the time of tendering or to be arrived at on the basis of a schedule of rates after completion, or it may be for an indeterminate amount arrived at on the basis of the actual cost of labour and materials used plus a percentage or fixed fee for establishment charges and profit. The offer may relate to a single job or to a series of jobs of a certain type and within a specified price-range over a period of years. However, as emphasised by McCanlis,[4] 'Any client who has a long term programme of work would be well advised in the interests of economy to plan his programme to achieve a steady flow of work commensurate with the resources of the selected contractors'. Where a building owner does not have a sufficient quantity of work to make this possible he is dependent upon the contractor's receiving work from other clients to provide the necessary continuity.

Implied terms

For small isolated jobs the contract may be verbal or contained in letters passing between the parties. In the absence of express conditions, the terms implied at common law are:

1. The building owner must allow the contractor to enter the building at the necessary time for the purpose of executing the work; he must give necessary instructions within a reasonable time and not obstruct the contractor in the performance of the work, and he must pay a reasonable price.
2. The contractor must do the work in a workmanlike manner and complete within a reasonable time.

In addition, where the contractor supplies materials there is an implied warranty that the materials are:

1. reasonably fit for the purposes for which they will be used, and
2. of good quality.

These common law obligations have been codified in the Sale

of Goods Act, 1979, which implies a condition that goods shall be reasonably fit for any purpose made known to the seller by the buyer, and the Supply of Goods and Services Act, 1982, which provides that, in the absence of express provision, anyone providing a service shall carry out the service with reasonable skill and care, within a reasonable time and at a reasonable price.

However, if the building owner does not rely upon the contractor's skill and judgment in choosing the materials the fitness warranty is excluded. This principle was laid down in Young and Marten *v.* McManus Childs Ltd 1968, in which the builder had been ordered to tile houses with 'Somerset 13' tiles which were apparently perfect on delivery but which later developed weathering defects necessitating their replacement. It was held that the builder would not have been responsible if tiles of good quality had been unsuitable for their purpose because the employer had chosen them, but he was liable for the defect in quality even though it was a latent defect undiscoverable by proper care on his part. In Gloucestershire County Council *v.* Richardson 1968 it was held that where the contractor is required by the building owner to obtain materials from a supplier on terms which severely restrict his right of action against the supplier, the quality warranty may also be excluded. The reason for this decision is that normally where the contractor is required to reimburse the employer in respect of defective materials he can recover the loss from the supplier who in turn may have a remedy against the manufacturer. However, where there is a collateral warranty the employer may sue the supplier direct. Thus in Shanklin Pier Ltd *v.* Detel Products 1951 the defendants warranted the suitability of their paint for repainting a pier in consideration of the plaintiffs' instructing the contractor to use the paint. In the event the paint proved a failure and the pier company recovered damages for breach of the warranty.

For contracts of any size it is highly desirable that the terms should be in writing and expressly stated to avoid any controversy at a later date over what was intended.

Usual conditions

The object should be to provide a fair and equitable legal

framework which will ensure that the work is carried out in a proper manner and that the contractor will receive a reasonable fee for his efforts. It is not intended to deal in detail with the legal relationships between the building owner and the contractor, but merely to point out those facets which have particular relevance in the context of planned maintenance. These are the related aspects of scope of work, price, time, quality, risk and disputes.

Scope of work

It is an essential requirement of any contract that there should be a clear understanding by both parties of their respective obligations. Thus the work which the contractor will be required to undertake should be as clearly defined as the circumstances permit. However, in some instances, full details will not be available at the time when the contractor is selected. This will obviously be the case where the contractor makes a standing offer to carry out work as and when ordered for a specified period. In such circumstances, while past experience will suggest the probable total workload for the period, the timing and extent of individual jobs will be uncertain. It will thus be necessary to agree a formula for arriving at a price for the work when details are available.

Usually the larger jobs will have been subjected to close scrutiny when preparing the budget estimates and the actual scope of the work determined by inspection. Where competitive tenders are to be invited the work will be programmed so as to allow sufficient time for the preparation of contract documents. However, there is no logical and systematic approach to the problems encountered in describing maintenance work. Methods vary from a detailed schedule of the individual items comprised in the work with the quantities of each item stated, to a broad general statement of the end result to be achieved.

The practical effect of these different methods lies in the amount of discretion allowed to the contractor. Where the work is stated in broad terms the onus is placed on the contractor to decide the precise extent of the work necessary and he is contractually liable to do all work indispensably necessary to achieve the end result. The detailed approach removes this discretion and, if the schedule is a contract document, the contractor is entitled to extra payment for work not specifically included therein. Clearly, where the work is described in broad terms variations in tenders will reflect not only differences in price levels but also different interpretations as to the amount of work involved. Where there is a large area of uncertainty there will be a natural tendency to inflate prices so as to cover every possible eventuality.

Where detailed information is given, it takes the form of separate statements about different aspects of the work which must be pieced together and usually supplemented by a site visit in order to obtain the complete picture.

The statements relate to:

1. Location
   Drawings.
   Locational references in the specification.
   Subdivision of the specification or the schedule of items according to location.

2. Work
   Description of operations.
   Extent of operations.
   Standard of workmanship.
   Restrictions – method/time.

3. Materials
   Type.
   Tests.

Much of the above is susceptible to standardisation; in particular it would be advantageous to standardise work descriptions and clauses relating to workmanship and materials. However, the traditional specification is too loose a form of communication and lacks the consistency necessary for this purpose. Attempts to speed up the process of producing specifications by standardising clauses have usually taken the form of comprehensive spot items which are incapable of adjustment to suit different circumstances.

Price

A major concern of the building owner is that the price he pays is fair and reasonable in relation to the amount and quality of work done and is within the budget estimates. The application of continuous cost control requires clear conditions relating to:

1. A statement of the lump sum or the formula which is to be used to arrive at the contract sum post-completion. Where the offer is in the form of a lump sum there should be an itemised breakdown of the work with each item separately priced for cost control and feedback purposes.
2. A statement of the documentary evidence that must be produced by the contractor as a prior condition to payment where work is post-priced, e.g. submission of certified time sheets and invoices in respect of materials for cost reimbursement contracts.
3. The periods at which interim payments will be made and the method by which they will be valued. For small jobs which can be completed within the space of one month interim payments are not necessary but for larger jobs there is an implied right to payment from time to time. Usually these payments would be made either at monthly intervals or when the contractor has reached an agreed stage in the work.
4. The procedures for ordering variations in the quality and quantity of the work and, in the case of lump sum contracts, the method of valuing such variations. For schedule and cost-reimbursement contracts, the contractor would be reimbursed automatically. A particular problem that might arise is whether or not the work is impliedly included. In many cases the descriptions are so loose that it is difficult to determine precisely what was intended. As a general rule if the work can be shown to be indispensably necessary to effect completion, it would be impliedly included even though not specifically described.
5. In addition, where the contract is of a cost-reimbursement type, there should be clear agreement as to what matters may be claimed as prime cost and what are deemed to be included in the percentage addition. Also it is desirable to give the client organisation greater control over the use of

resources than would be appropriate for other types of contract. This control would extend to the method of working and sequence of operations, the number of men employed on the site, the purchase of materials and the plant used on the site.

Time

The timing of the larger items of work will have been fixed when drawing up the master programme, with the object of causing as little interference as possible with user activities and also to ensure that payments for work done phase in with the overall cash flow pattern of the firm. It is probable therefore that failure to complete by the specified time could result in losses to the firm. To cover this contingency, it is usual to provide for the contractor to pay the building owner an agreed amount for each week of delay beyond the date of completion. However, it is important that the amount of liquidated damages should be a genuine pre-estimate of the actual loss likely to be suffered, otherwise it may be construed as a penalty and be unenforceable. Also, there should be a parallel clause empowering the building owner to extend the time for delay caused by specified events beyond the control of the contractor. Otherwise, the ordering of additional work by the building owner would render the original date of completion inappropriate and, in the absence of provisions for the substitution of an extended date of completion, the building owner would lose his right to claim the damages.

Where the agreement is to carry out work as and when ordered, it is not possible to lay down commencement and completion dates at the time of entering into the contract. Such dates could be given when ordering the individual jobs but reliance would have to be placed on the contractor's obligation to complete within a reasonable time. The building owner may be given a contractual right to determine the contract if he considers that progress is not satisfactory, but in practice such a power should be exercised with caution to avoid a claim for wrongful dismissal.

Quality

The contract should contain appropriate safeguards to ensure that the quality of the work done is satisfactory. A primary obligation into which the contractor is normally required to enter is to 'carry out and complete the works to the reasonable satisfaction of the building owner or his representative'. What is reasonable would depend on the surrounding circumstances and, in particular, on the price paid.

Conditions which seek to safeguard the employer's interests in this respect are of two types: those which attempt to prescribe the circumstances which will favour the production of a reasonable quality, e.g. a requirement to keep a competent foreman on the site at all times, and those which provide a building owner with a remedy if the quality is unsatisfactory, e.g. contractor to make good any defects due to non-compliance with the specification for an agreed period following completion of the works.

Additionally, where there is no final and conclusive certificate the contractor's liability for work not in accordance with the contract extends for a period of six years in the case of a simple contract and twelve years if the contract is by deed from the date of the cause of action as provided in the Limitation Act, 1980. Where fraud can be imputed, the foregoing periods would run from the date of discovering the fraud.

Risk

The execution of building work inevitably involves some risk of personal injury or damage to property. It is essential that such risks should be adequately covered by insurance, preferably by a joint names policy to cover all third-party liability irrespective of who is negligent. In the case of maintenance, the existing building would normally be insured against damage by fire and other common risks and additional cover could be arranged for the extra risk created by the building operations.

While it is a general rule of law that a person is not responsible for the torts of an independent contractor, there are exceptions which could be important in the context of maintenance.

Thus, the building owner may be liable:

1. if the work causes loss to a third party and the employer has not imposed a duty of avoiding such loss on the contractor;
2. where the work of its very nature involves a risk of damage to a third party, e.g. interference with the right of support of an adjoining building;
3. where fire is negligently caused on the site by the contractor in the performance of the contract and it spreads, causing damage to adjoining buildings.

These areas of strict liability should be covered by insurance rather than relying on an indemnity from the contractor which in the event may prove valueless.

In some circumstances, the responsibilities of the occupier and the contractor may overlap, e.g. under the Construction Regulations the contractor must provide safe means of access for his men, while under the Factories Act the occupier must provide safe means of access for every person employed or working on the premises and this would include the contractor's men. Thus, if safe means of access are not provided and as a result one of the contractor's men is injured, the occupier and the contractor would be jointly liable. It is, therefore, incumbent upon the client's maintenance organisation to ensure that all necessary precautions are taken.

Another type of risk which may have to be guarded against applies where the workmen may gain information on processes which would be of interest to the firm's competitors, or which may involve matters of national security.

Disputes

Many maintenance contracts are so loosely worded and the scope of the work so ill-defined that the smooth running of the contract calls for a good deal of give-and-take on both sides. It is perhaps only the smallness of the sums involved that dissuades the parties from engaging in costly litigation. For jobs of any size it is desirable to lay down procedures for dealing with any disputes that may arise, usually by providing that they should be referred to a named arbitrator or one who is to be selected in an agreed way. This method of settling disputes is generally regarded as quicker and cheaper than the ordinary processes of the law.

Contract documents

The contract documents will depend upon the size, nature and complexity of the work and whether payment is to be made on a lump sum, schedule or cost reimbursement basis. They may include a separate form of agreement with attached conditions, drawings, specifications, bills of quantities and schedules of rates.

Agreement and conditions

It is advantageous to use a standard form of agreement wherever possible. Unfortunately there are few standard forms available that cover the types of contractual arrangements required for the general run of maintenance work. The principal standard forms are:

JCT Standard Form of Building Contract, 1980 edition
This form is issued by the Joint Contracts Tribunal which is a committee composed of representatives of the RIBA, BEC, RICS and various local authority and sub-contractors' associations. There are three variants of the Form – With Quantities, Without Quantities, and With Approximate Quantities – which is published in two editions, one for Private Use and the other for Local Authority Use. In addition the JCT have issued a complementary set of documents for nominated sub-contracts – a standard form of tender, employer/nominated sub-contractor agreement, form of nomination and sub-contract agreement and conditions. The provisions are exceedingly complex and have been designed to cover the requirements of new construction works of some magnitude.

Agreement for Minor Building Works (see Appendix C)
This form is also issued by the JCT for minor building works

for which a lump sum has been agreed and where an architect/supervising officer has been appointed on behalf of the employer. The heading to the form states that it is not suitable for use where a bill of quantities has been prepared, or where the employer wishes to nominate sub-contractors or where the works or services are of a complex nature. However, in spite of its title, the form can be used for quite large jobs where the extent of the work to be done is reasonably certain and can be clearly defined so as to provide a realistic basis for a lump-sum tender.

JCT Intermediate Form of Contract (1984)

This form is issued for contracts in the range between those for which the JCT Standard Form and the Minor Works Agreement are used and is suitable where the works involve the normal basic trades without complex specialist services. It requires the employer to provide at tender stage a set of drawings together with another document. Where this is a bill of quantities or schedule of work the contractor is required to price it but where a specification he may supply a contract sum analysis. The conditions are less detailed than those in the Standard Form and in particular do not contain the complex provisions relating to the appointment of nominated sub-contractors. Instead provision is made for work which is to be priced by the contractor to be carried out by a person named in the supporting tender document.

Guidance on the use of the various JCT Forms is given in JCT Practice Note 20 (Revised 1984), which suggests that subject to the complexity of the work the Intermediate Form should be suitable for contracts up to £250,000 (1983 prices) and the Minor Works Agreement for contracts up to £50,000 (1981 prices).

Fixed Fee Form of Prime Cost Contract

This is based on the Standard Form of Building Contract and requires the prior preparation of drawings and/or a specification on which the contractor bases his fixed fee for overheads and profit. It envisages new construction on a more extensive scale than most maintenance works and assumes that there is sufficient knowledge of the scope of the work at the time of tendering to produce a realistic estimate on which the contractor can base his fixed fee. On completion the contractor is paid the prime cost of labour, materials and plant as defined in the contract plus the fixed fee for overheads and profit.

Model Conditions of Contract for use with the National Schedule of Rates

This is modelled on the form that was produced by the BEC for local authority measured term contracts. It differs significantly from the JCT forms of contract which envisage single lump sum contracts. In particular it provides for a contract area and a contract period to be stated in the tender and for determination without default by either party at the end of six months from the date of commencement and thereafter on giving six weeks' notice. Orders for individual jobs during the contract period are to be in writing and should state a reasonable date for completion. The supervising officer is made responsible for arranging access to premises and the contractor entitled to recover the value of unproductive time should he be unable to gain access at the requisite time. Work is to be valued in accordance with the National Schedule of Rates or analogous rates or, where this would be unreasonable, as daywork in accordance with the *Definition of Prime Cost of Building Works of a Jobbing or Maintenance Character* published by the RICS/NFBTE. The contractor is required to prepare his accounts for each and every order within 28 days of completion of the work which is to be certified by the supervising officer within 28 days of receipt and payment made within 14 days of certification. Any defects that appear within three months of completion are to be made good by the contractor. Other provisions such as insurance against personal injury and damage to property are similar to those in the JCT Minor Works Agreement.

Drawings

The main way of conveying construction information to the contractor is by means of annotated drawings drawn to appropriate scales. The Building Regulations require the following scales to be used for deposited plans – 'plans and sections shall be to a scale of not less than 1:100 or, if the building is so extensive as to render a smaller scale necessary, not less than 1:200; block plans shall be to a scale of not less than 1:1250; and key plans to a scale of not less than 1:2500'. Block plans are described as those showing the size and position of the building and its relationship to adjoining buildings and a key plan as one which shows the position of the site when it is not sufficiently identifiable from the block plan. In addition larger scale details are often necessary to illustrate complex or critical parts of the structure. The standard conventions for the presentation of drawings are given in BS 1192 (Building Drawing Practice).

Specification

The presentation of information in a specification may take different forms according to the use of the specification and whether or not it is to be a contract document. For the larger new works a specification may be prepared by the architect as part of the initial brief to the quantity surveyor to assist in the preparation of bills of quantities. For smaller new works such as alterations and extensions it is unlikely that there will be a bill of quantities and the contractor will be required to base his tender on the specification and drawings. For some types of maintenance work drawings may be unnecessary and the specification will be the sole source of information concerning the work to be done. In all cases, however, the information should be clear and unambiguous. Also care should be taken to ensure that statements about such matters as the nature of the sub-soil are correct otherwise the contractor may be able to rescind the contract or claim damages under the Misrepresentation Act, 1967, or, if the statement has become a term of the contract, sue for breach of contract. Assistance in drafting clauses may be obtained from the National Building Specification or from the publication Specification.

A specification is normally divided into two main parts – Preliminaries and Materials and Workmanship.

Preliminaries
This section gives provisions which govern the general con-

duct of the contract and the overall extent of the contractor's liabilities. The clauses will have to be drafted to meet the requirements of the particular job and will include such matters as:

General description of work.
Form of contract to be used.
Provision of plant, scaffolding, etc.
Provision of sheds for storing cement, etc.
Office for Clerk of Works (if there is one).
Water, lighting and power.
Protective measures where work in occupied buildings.
Times of access to site, restrictions on method of carrying out work and other items likely to affect progress.

Materials and workmanship
This section describes the quality of the materials to be used and the method of construction and standard of workmanship. For new works it is convenient to follow the order in which the Work Sections are given in the Standard Method of Measurement of Building Works (6th edition).

Demolition
Excavation and earthwork
Piling
Concrete work
Brickwork and blockwork
Underpinning
Rubble walling and masonry
Asphalt work
Roofing
Woodwork (carpentry and joinery)
Structural steelwork
Metal work
Plumbing and mechanical engineering installations
Electrical installations
Floor, wall and ceiling finishings
Glazing
Painting and decorating
Drainage
Fencing

Of course, the above list will have to be modified according to the nature and size of the job. For small works many of the work sections will not be required and also it may be advantageous to group two or more work sections together, e.g. Excavation and Concrete work.

For works in alterations and repairs it will often be necessary to depart from the sequence of Work Sections and adopt a sequence which follows the order in which the work will be carried out on site. In other cases such as internal decoration it may be more convenient, both from the point of view of preparing the specification and pricing, to group the work according to rooms in the order in which the rooms would be visited when walking round the building.

However, assuming that the Work Section order is to be followed, the clauses within each work section would be grouped as follows:

(*a*) Clauses of general applicability to the work section.
(*b*) Materials and their preparation, application and protection (in some cases it may be better to group all the materials in a separate section to avoid repeating the description in each work section where the material is used).
(*c*) Work to be done.

The materials are described in the following ways:

(*a*) Giving a full description stating desirable and undesirable features and any tests with which they should comply.
(*b*) Stating the relevant British Standard Specification – note, however, that the BSS is usually the minimum quality and that in some cases more than one quality is provided for.
(*c*) Specifying a proprietary brand or naming a particular manufacturer or source of supply.
(*d*) Giving a prime cost (Pc) sum and an outline description of the material – useful for such items as sanitary fittings in that decisions as to colour, etc. can be left until nearer the time of fixing.

Adjectives such as 'best' or 'first class' should be avoided unless they are recognised terms used to describe the particular quality required. Also the term 'other equal and approved' should be treated with caution.

The workmanship clauses usually follow the order of carrying out the work on-site and should state precisely what is to be done and how, giving details of any constraints on the method of working, e.g. concreting to stop when the temperature falls below a certain level. Use can be made of BS Codes of Practice where appropriate and standard specification clauses for work of a repetitive nature. The right to select subcontractors for specialist work may be reserved either by naming the subcontractor or by including a prime cost sum to cover the cost of carrying out the work. Where a part of the work is too ill-defined to describe accurately at the time of preparing the specification a provisonal sum may be included to cover the estimated cost and an adjustment made to the contract sum on completion.

The type of specification described is called an operational specification in that it describes the actual physical work to be carried out by the builder. An alternative is the performance specification which specifies the performance to be achieved and leaves it to the builder to use his skill and knowledge to devise a physical solution which will meet the performance requirements. Thus, in the case of a central heating system one could state that the system is required to maintain a minimum temperature of 21°C in each room when the outside temperature is −1°C and leave it to the contractor to design the system. This places a greater liability on the contractor in that he is liable not only for the quality of the materials used and the standard of workmanship but also for the fitness for purpose of the system, i.e. that it meets the performance requirements.

Bills of quantities

Bills of quantities are normally only used for the larger jobs and where used in conjunction with the JCT Standard Form of Contract must be prepared in accordance with the Standard Method of Measurement of Building Works issued jointly by the RICS and the BEC. A bill of quantities sets out in a systematic manner the quantities and full descriptions of all the items of labour, materials and plant required to erect and complete a building. The Preliminary section of the bill gives a brief description of the works and of the type of contract to be used and also details of general matters which may affect the price of the job. The measured work which follows is grouped into appropriate sections and each section starts with a set of preamble clauses which describe the quality of the main materials to be used and the standard of workmanship. Alternatively instead of giving the preamble clauses at the beginning of each section they may be grouped together in a separate section.

Usually the Work Sections follow the order in which they are given in the Standard Method of Measurement but other groupings sometimes adopted are by functional elements, i.e. parts of the building which always perform the same function irrespective of the type of construction, or by operations, i.e. site activities.

The bill of quantities has two main purposes – firstly, to provide a uniform basis for competitive tenders by providing each contractor with an accurate statement of the quality and quantity of the work in a form suitable for pricing, and, secondly, to provide a schedule of prices for the valuing of variations ordered during the course of the contract. In addition the priced bills facilitate the valuation of interim certificates and the settlement of the final account and provide useful cost information for planning future projects.

Schedules of rates

These may be *ad hoc* schedules prepared for a particular job or standard printed schedules for use in term contracts. The *ad hoc* schedules usually follow the same pattern as a bill of quantities but may be either without quantities or with approximate quantities to assist in the evaluation of tenders. A standard schedule is one which has been designed to cover a range of repetitive jobs and may be one which is peculiar to a particular organisation or one which is more broadly based and of general applicability.

The Property Services Agency of the DOE produces separate standard Schedules of Rates for Building Works, for Minor Building Works and for Decoration Work. These are used mainly for term contracts and form the basis for estimating, tendering and pricing. More recently a National

Schedule of Rates has been produced jointly by the Building Employers Confederation and the Society of Chief Quantity Surveyors in Local Government.[5] It is a comprehensive microcomputer-based system which has been designed principally for housing work with a maintenance bias. Each item in the schedule is broken down into labour and material elements so that it may also form the basis for bonus payments and materials allocation. The rates are updated quarterly to reflect the effect of inflation and changes in the pattern of prices. Also, the BMCIS Price Book provides a useful schedule of the more common maintenance items and the rates are similarly broken down into their labour and material components.

The term Schedule of Rates is also used to describe the list of rates which have been used by the contractor in building up his tender for a lump sum contract and which he is required to submit for the purpose of valuing any variations that might be ordered.

References

[1] Department of the Environment, Property Services Agency (1980) *Schedule of Rates for Building Works/Minor Building Works and Maintenance of Roads and Pavings/Decoration Work*. HMSO.
[2] Research Institute for Consumer Affairs (1971) *The Consumers' View of Building Maintenance*. Department of the Environment.
[3] Orchard, R. R. (1966) 'The execution of maintenance – the contractor.' Edinburgh: Conference on Building Maintenance.
[4] McCanlis, E. W. (1967) *Tendering Procedures and Contractual Arrangements*. Research Report, RICS.
[5] The National Schedule of Rates (SCQSLG/BEC). OCM Publishing.

CHAPTER TEN

Diagnosis of building defects

Causes of defects

A building defect may be defined as an imperfection, deficiency or fault in a building element or component which adversely affects its functional performance or appearance. Some defects are the natural consequence of ageing and normal use but many premature failures can be traced to a lack of proper skill and care. A BRE study[1] concluded that over 90 per cent of building defects could be explained partly or wholly as the result of readily identifiable faults in design or construction which could have been foreseen and hence prevented.

Broadly, the causes may be attributed to the following:

1. Inadequate brief. It is often said that defects start on the drawing board but in some cases they can originate at an even earlier stage. For example, the brief may lay down totally unrealistic cost limits or fail to give vital information on the functional requirements of the building. Usually there is no indication of the likely period of use nor of the client's attitude towards maintenance.
2. Faulty design decisions. The most common faults may be grouped as follows:

    (a) Failure to follow well established design criteria in the choice of structural system and selection of materials.
    (b) Ignorance of the basic physical properties of materials, e.g. failing to make allowance for the differing thermal

and moisture movements of materials used in combination.

(c) Use of new materials or innovative forms of construction which have not been properly tested in use. This is often the result of uncritical reliance on manufacturers' literature quoting simulated laboratory tests.

(d) Misjudgement of user and climatic conditions under which the material will have to perform.

(e) Complex details which have a low probability of successful execution on an open building site.

(f) Poor communications between different members of the design and construction teams.

The fault may be traceable to component manufacturers, specialist sub-contractors and consultants as well as the main designer.

A less obvious design fault is the failure to consider the ease with which components can be maintained and eventually replaced. For example, little thought is given to the standardisation of components in order to reduce the need to carry a large variety of spare parts or to ensuring that access can be easily gained for servicing and cleaning.

3. Construction methods. The conditions under which construction takes place are often far from ideal and, coupled with an emphasis on speedy completion, can result in careless and skimped work. Although the BRE study[1] mentioned earlier showed that only a small proportion of defects were attributable to faulty materials it is apparent that some manufacturers of so-called high technology components have little awareness of the rigours of a building site or the standards of accuracy achievable under such conditions. Thus, whilst the materials may be perfect on leaving the factory they can quite easily be damaged in transit, loading and unloading, unsuitable conditions of storage on site and hoisting and placing in position. Many such defects could be avoided by ensuring greater care at all stages in the process, proper training of operatives, and closer supervision. To tackle this problem the construction industry is beginning to introduce the quality assurance techniques developed in other industries such as Quality Assurance (QA) groups and circles (QC). Essentially these

techniques consist of setting down appropriate inspection procedures and specifying levels of acceptance and rejection together with methods of sampling and testing the performance characteristics.

4. User activities. Defects may be caused by unintentional misuse through a lack of information on the correct mode of use, or by deliberate acts of vandalism. The solution is to provide the designer with more information on the degree of severity of use so that a better match can be made between the robustness of the fittings and finishings and the conditions of use. Also, certain defects may be related to the social attitudes and financial circumstances of the user, e.g. condensation is affected by the amount of money spent on heating and ventilating, and the occupancy pattern.

5. Maintenance. Incorrect identification of the true cause of a defect, and inappropriate remedial work, will not only do nothing to rectify the original defect but may substantially worsen the condition of the building. Similarly, lack of care in carrying out repairs and inspections may be the cause of defects in previously satisfactory elements, e.g. walking on unprotected felted flat roofs can drive the gravel into the felt, causing splits and cuts leading to premature leaking. The life of building elements and components can be extended considerably by adopting a planned maintenance approach so that problems can be identified in their early stages and preventive maintenance carried out to avoid early failure.

Diagnosis of defects

The general approach should be to examine carefully all the symptoms, consider all the probable causes, by a process of elimination identify the true cause and its source, and then decide on appropriate remedial action. In many instances a visual examination will be sufficient for an experienced inspector to determine the cause. Where this is not possible instruments may be used for a more objective diagnosis. There are many portable instruments on the market for investigating the causes of defects, including:

> *Moisture meter*, to test the comparative moisture content of materials.

Ultrasonic tester, for checking concrete strength.

Depth meter, to enable the cover to reinforced concrete to be ascertained. This instrument can also be used to locate the presence of metal pipes in unreinforced concrete.

Endoscope, which comprises a slender tube fitted with a magnifier and light that can be inserted through a small hole to give a view of the interior of wall cavities and other voids within a building.

In some cases it may be necessary to monitor the rate of development of a defect over a prolonged period in order to assess the probability of failure. Examples include measuring the width of cracks by means of pins and a micrometer, taking levels at fixed points to determine the rate of settlement and checking vertical alignments with the aid of a theodolite.

Where the surface manifestation of a defect is not sufficient to indicate the true cause, and particularly where there is a reasonable possibility that a structural defect is concealed, an exploratory survey should be undertaken. Such investigations should be carried out carefully to minimise consequential damage, and can be facilitated if the design allows ready access to hidden parts of the structure, e.g. strategically placed access panels to ducts and ceiling hatches. Specialist advice may be required in respect of certain defects, e.g. structural stability and services and laboratory tests of subsoils or materials to establish their composition and performance characteristics.

The following tables list the probable causes of the more common defects but for more detailed information the reader should refer to the relevant BRE Digests.[2,3]

Tables of defects and causes

FOUNDATIONS AND BASEMENTS		
Element	Defect	Cause
Foundations	Settlement causing cracks in supported walls	Overloading through (i) inadequate design, e.g. width or depth insufficient in relation to bearing capacity of subsoil

	FOUNDATIONS AND BASEMENTS	
Element	Defect	Cause
Foundations (*contd.*)		(ii) building used for a purpose different from that for which it was designed
		(iii) structural alterations, e.g. formation of large openings in supported wall concentrating load on adjacent foundations
		(iv) building additional storey without strengthening existing foundations
		Shrinkage of clay subsoils through withdrawal of ground water by prolonged drought, nearby fast growing trees such as poplars, site drainage, or tunnelling
		Undermining of foundations as a result of finer particles of sandy soils being washed out by underground streams, flooding or leaky drains
		Long-term settlement of peat and made ground
		Downhill creep of clay soils
		Swallow holes and mine workings under foundations
		New excavations immediately adjacent to building to greater depth than old foundations
		Foundations of extensions, bays, screen

Foundations and Basements

Element	Defect	Cause
Foundations (*contd.*)		walls, etc., with shallower foundations than those to the main building
		Vibrations from heavy traffic, machinery, piling operations, tunnelling, etc.
	Heave causing lifting and cracking of ground slabs	Freezing of water in upper layers of soil
		Expansion of clay soils on absorbing moisture during prolonged wet weather or as a result of felling large mature trees close to the building
	Disintegration of concrete	Sulphate attack – reaction between tricalcium aluminates in cement and soluble sulphates in soil
Basements	Dampness	Defective tanking through (i) deterioration with age (ii) lack of continuity, e.g. not carried under stanchions and other structural members which penetrate the tanking (iii) damage occurring during construction
		Opening up of construction joints due to shrinkage of concrete where dense concrete used instead of tanking
		External water pressure caused by raising of water table adjacent to building

Element	WALLS AND FACINGS Defect	Cause
External walls	Diagonal cracks following horizontal and vertical joints	Settlement of foundations
	Horizontal and vertical cracks	Thermal movement due to omission of expansion joints in long lengths of walling
		Moisture movement due to expansion of bricks used too soon after firing
	Outward bulging of wall between ground level and eaves with gap internally between wall and edge of upper floors	Displacement of wall through one or more of the following: (i) low slenderness ratio, i.e. thickness of wall insufficient in relation to height (ii) inadequate lateral support, e.g. cross walls or floors not built into wall to restrain movement (iii) overloading of structure (iv) vibrations from heavy traffic or machines
	Outward bulging of wall at top	Spreading of untied rafter feet exerting horizontal thrust on supporting walls
	Horizontal cracking of inner leaf of cavity wall	Sulphate attack causing expansion of mortar joints in outer leaf
	Bowing of external leaf of cavity wall	Insufficient wall ties or corrosion of poor quality ties
	Oversailing of outer leaf of cavity wall over damp-proof course	

Element	WALLS AND FACINGS Defect	Cause
External walls (*contd.*)	General internal surface dampness	Condensation through inadequate heating or ventilation
	Horizontal tide mark staining decorations above ground floor level	Upward penetration of damp through: (i) absence of dpc (ii) defective dpc (iii) bridging of dpc by flower beds, raised paths, mortar droppings in cavity, floor screed, etc.
	Isolated damp patches	Lateral penetration through: (i) solid walls (ii) mortar droppings on wall ties of cavity walls (iii) faulty cavity filling Downward penetration through: (i) defective roof coverings, gutters or flashings (ii) omission of dpc under coping to parapet or cavity tray at roof level (iii) omission of dpc to chimney stack or defective gutter or flashings
Openings	Cracks in soffit	Faulty lintel: (i) inadequate size (ii) concrete lintel upside down with reinforcement at top instead of bottom (iii) decayed timber lintel in older property

Element	WALLS AND FACINGS Defect	Cause
Openings (*contd.*)	Damp patch over head of opening	Omission of cavity tray over opening in cavity wall or tray too short if damp patches in corners only
	Damp patches to reveals	Omission of vertical dpc at junction of inner and outer leaves or window/door frame set too far forward
	Damp stains at sill level	Condensation running down inner face of glass
External Facings (a) Brick	White powdery deposit	Efflorescence caused by crystallisation of soluble salts in bricks or mortar
	Spalling of surface	Sulphate attack – crystallisation of salts in outer layers of friable bricks
		Frost action – expansion of absorbed water on freezing
		Weathering of the less durable under-burnt bricks
	Cracks in calcium silicate brick outer leaf used with fletton inner leaf	Differential movement – drying shrinkage of calcium silicate bricks accompanied by expansion of flettons
	Bowing of brick facings to concrete wall	Differential movement – drying shrinkage of concrete with possible moisture expansion of freshly fired bricks

WALLS AND FACINGS

Element	Defect	Cause
External Facings (a) Brick (*contd.*)	Staining of bricks	Run-off from concrete roof or external stairs or washings from limestone features
		Leaking gutters, cracked rainwater pipes or overflows with inadequate projection
(b) Stone	Pitting and staining of limestone plinths	Rising damp – pavement finished at a higher level than wall dpc or, where there is a basement, lack of continuity of wall dpc and tanking
	Decay of sandstone plinths	Washing from limestone wall above
	Unsightly streaks of dirt	Uneven washing of dirt deposits through inadequate projection of copings and sills and omission of traditional string courses
	Delamination of surface layers	Improper laying with respect to quarry bed allowing penetration of rainwater with subsequent disruption of surface on freezing
		Sulphate attack from soluble salts in backing or bedding mortar
	Erosion of surface	Wind blown grit
	Blistering of surface	Atmospheric pollution – sulphur-dioxide absorbed into rainwater to form a weak acid solution which attacks the binding medium of the stone

Element	WALLS AND FACINGS Defect	Cause
External Facings (b) Stone (*contd.*)	Isolated brown or green stains on surface	Run-off from metal fixings cramps
	Displacement of thin stone facing slabs on brick backing	Rusting of ferrous fixing clamps and anchors
		Sulphate attack – soluble salts in backing or bedding mortar
		Omission of movement joints
(c) Slate	Bowing of slate facing slabs from concrete backing	Differential thermal movement of slate and concrete – although having much the same coefficient of thermal expansion the slate, being dark and exposed to direct sun, becomes much hotter than the backing concrete
(d) Rendering	Map pattern cracking	Drying shrinkage of dense cement and sand mix
	Loose patches	Inadequate key to backing or freezing of entrapped water
	Horizontal and vertical cracks	Sulphate attack – rainwater admitted through shrinkage cracks of dense mortar providing vehicle for sulphates in bricks to react with cement (possibly use of gypsum plaster instead of lime in gauged mortar being source of sulphates)
		Structural movement due to settlement or to thermal or moisture movement of backing

Element	WALLS AND FACINGS Defect	Cause
Parapets	Cracks in rendering to parapet	Sulphate attack – omission of dpc under coping allowing penetration of rainwater into body of wall where it provides a vehicle for soluble sulphates in bricks to react with cement in rendering
	Parapet to flat roof moved outwards over supporting wall	Thermal and moisture movement of concrete paving slabs on roof – under cold dry conditions the slabs shrink, opening up the joints which fill with debris so that they cannot close when the slabs expand under warm and wet conditions
Stacks and flues	Dampness in upper part of stack	Defective dpc, flashings or chimney gutter
	Smokiness	Insufficient draught due to poor flue design, overlarge throat, partial blockage of flue or stack not high enough
	Leaning stack	Sulphate attack causing expansion of mortar joints. As the north facing side is not exposed to the drying influence of the sun the joints on this side will expand more than those on the south side causing the stack to lean over to the south
	Damp patches and discoloured decorations on chimney breast	Condensation from flue gases through using unlined flues servicing slow burning appliances

Element	WALLS AND FACINGS Defect	Cause
Stacks and flues (*contd.*)		Often occurs after changing to slow burning fuels after many years using coal
		May also occur where the fire opening has been sealed but the top of the flue left open. Rain may enter the top of the flue and with no heat to dry it out eventually causes damp patches with stains from the chlorides derived from the soot
Partitions	Cracks at head and sides	Drying shrinkage of lightweight concrete blocks
	General cracks	Inadequate support at base: (i) settlement of foundations of ground floor partitions (ii) deflection of upper floor joists under load (iii) new partition built on old floor of insufficient strength
		Creep of concrete floor
		Differential movement between partition and surrounding structure

Element	FRAMES AND CLADDINGS Defect	Cause
Frame (a) Timber	Rot	Rising damp – omission or penetration of dpc under sill

FRAMES AND CLADDINGS

Element	Defect	Cause
Frame (a) Timber (*contd.*)		Direct penetration of rainwater – cladding not weatherproof or omission of water barrier
		Interstitial condensation – omission of or torn vapour barrier on warm side of insulation
		Initial protection of visible surfaces only, together with inadequate maintenance
(b) Steel	Wave distortion of frame	Unequal settlement of columns, excessive loading or thermal movement
	Cracking and rust staining of concrete cover	Corrosion of steel – cover concrete of inferior quality or inadequate thickness, mechanical damage of exposed angles
(c) Concrete	Cracking and disintegration of concrete	Weak concrete due to improper mix design, unsuitable aggregates or cement, wrong water/cement ratio, inadequate compaction and curing
		Secondary effect of freezing of water which enters cracks causing spalling of surface layers of concrete
		Use of calcium chloride as an additive to accelerate setting of concrete
		Insufficient cover to reinforcement – increased loss of alkalinity caused by carbonation of

<table>
<tr><td colspan="3" align="center">*FRAMES AND CLADDINGS*</td></tr>
<tr><td>Element</td><td align="center">Defect</td><td align="center">Cause</td></tr>
</table>

Element	Defect	Cause
Frame (c) Concrete (*contd.*)		concrete causes steel to rust and build up sufficient pressure to crack thin concrete cover
		Overstressing caused by compressive, tensile or shearing forces
	Brown stains	Pyrite staining – commonly found with Thames gravel aggregates containing iron sulphide which reacts with water and oxygen to form brown iron hydroxide
		Rusting of reinforcement through penetration of water where cover of weak porous concrete or inadequate thickness
	Blistering and peeling surface	Effect of polluted atmosphere containing sulphurous gas on poor quality concrete
		Effect of sea water and marine atmospheres on exposed concrete
	Pattern staining and efflorescence on fair faced concrete	Formwork – use of different quality boards or mixture of old and new boards causing uneven attachment of concrete and irregular curing
		Excess water from freshly placed concrete running over surface of concrete underneath
	Sagging and cracking of beams	Inadequate design and/or workmanship

Element	Defect	Cause
FRAMES AND CLADDINGS		
Frames (c) Concrete (*contd.*)		Overloading
		Use of high alumina cement
Claddings (a) Weather-boarding	Splitting and cupping of boards	Boards too thin
		Lack of maintenance – possibly mistaken view that teak and red cedar need no maintenance
	Rot	Water entering end grain at top or bottom of vertical boarding through absence of proper flashings and throatings
		Horizontal t & g boarding laid with grooves uppermost
		Interstitial condensation through lack of vapour barrier and ventilation cavity
(b) Tile hanging	Isolated tiles missing	Accidental breakage or wind suction on badly fixed tile
	Substantial number of missing tiles	Rot in battens or corrosion of nails
	Bottom edge uneven	Rot in tilting fillet
(c) Brick panels	Outward bowing of panels	Differential movement between concrete frame and brickwork
(d) Concrete panels	Uneven weathering and dirt streaks	Inadequate projections to throw rainwater clear of surface or run-off from ends of sills

FRAMES AND CLADDINGS

Element	Defect	Cause
Claddings (d) Concrete panels (*contd.*)	Rust stains	Inadequate cover to reinforcement or poor quality porous concrete
	Surface disintegration	Expansion on freezing of water which has entered cracks or pores in concrete
	Damp penetration	Badly designed joints or poor quality sealants
Curtain walling (a) Timber	Deformation	Seasonal movement of timber
	Damp penetration	Gap between frame and glass too small to accept differential movement – omission of spacers
		Defective sealants incapable of accommodating differential movement of materials making up system
		Failure to provide a drained cavity where back-up panel used
(b) Aluminium	Bulging between columns	Omission of movement joints

ROOFS AND COVERINGS

Element	Defect	Cause
Construction (a) Timber	Sagging ridge and rafters	Timber sections too small or improperly framed and strutted

ROOFS AND COVERINGS

Element	Defect	Cause
Construction (a) Timber (*contd.*)		Removal of vital supports during structural alterations, e.g. purlins cut to facilitate access to an attic
		Lack of longitudinal ties to trussed rafter roofs
		Overloading of roof structure by replacement of light covering such as slates with heavy concrete tiles
		Corrosion of metal fixings such as nails, bolts and gangplates, possibly due to use of timber preservatives based on metal salts causing bi-metallic corrosion
	Rot in structural timbers	Wet rot caused by ingress of water through defective coverings or condensation
		Dry rot where moisture content of timber in badly ventilated positions exceeds 20% – usually migrates from other less accessible parts of the building via joints in brickwork
	Rot in fascia and ends of rafters	Wet rot – top of cavity wall not sealed
	Insect	Usually the common furniture beetle but in parts of Surrey larger flight holes might indicate infestation by the more destructive house

	ROOFS AND COVERINGS	
Element	**Defect**	**Cause**
Construction (a) Timber (*contd.*)		longhorn beetle. The death watch beetle is associated with oak in old buildings and the powder post beetle with certain hardwoods which have a higher sugar content
(b) Concrete	Cracks along top of supporting wall	Thermal expansion of concrete
(c) Steel	Corrosion	Water penetration through coverings or condensation
		Lack of paint or other protective treatment – perhaps through sections being too close together to permit all-round application
Coverings (a) Slates and tiles	Isolated slates or tiles missing	Accidental breakage or wind suction
	Aeas of slates or tiles missing or slipped	Rot in battens or corroded nails
	Slipped hip tiles	Missing hip irons
	Lamination of tiles	Frost action – porous tiles laid to too flat a slope
(b) Built-up felt	Tearing	Poor quality felt or differential movement or cracks in substrate
	Blistering	Expansion in warm sunny weather of entrapped air or water
	Cockling and rippling	Moisture movement of felt or roof deck

ROOFS AND COVERINGS

Element	Defect	Cause
Coverings (b) Built-up felt (*contd.*)	Lifting of laps	Faulty initial bonding
	General deterioration and embrittlement	Age, traffic, inadequate falls, incorrect choice of felt or method of laying, solar radiation causing thermal stress
(c) Asphalt	Ponding	Localised depressions in surfaces laid flat or to shallow falls
	Cracking	Differential movement of asphalt and base (particularly pre-cast concrete slabs)
	Cracking at junction with skirting	Asphalt unable to accommodate movement or deflection of base where skirting rigidly fixed to parapet wall
	Blistering	Expansion in warm sunny weather or entrapped air or moisture
	Crazing and general surface deterioration	Age, traffic, lack of protective or reflective covering to reduce solar radiation
(d) Metal (lead, copper, aluminium, zinc)	Damp penetration	Rainwater entering by capillary action through improperly formed rolls, seams, etc., or defective flashings
	Tears and cracks	Movement of base
		No provision made for thermal movement
	Pitting and embrittlement	Electrolytic action between dissimilar

ROOFS AND COVERINGS		
Element	Defect	Cause
Coverings (d) Metal (lead, copper, aluminium, zinc) (*contd.*)		metals used in juxtaposition
	Small circular holes	Electrolytic action between covering and nail heads used for fixing roof boarding
(e) Sheet (asbestos cement, corrugated iron)	Damp penetration	Inadequate laps or nailed through troughs instead of crowns of corrugations
	Embrittlement of asbestos cement sheets	Ageing, industrial pollutants
	Premature rusting of galvanised iron sheeting	Protective zinc coating damaged – perhaps as a result of punching instead of drilling holes
Roof drainage (a) Gutters	Irregular fall	Rot in fascia or insufficient brackets
	Rust or dirt runs on underside of cast-iron gutters	Leaking joints
	Cracking or sagging of plastic gutters	Inadequate provision for thermal movement
(b) Rainwater pipes	Out of line	Defective or missing fixings
	Rust on back of cast-iron pipes	Clearance between pipe and wall insufficient to allow proper paint coverage
	Blockage of pipe	Lack of wire ballon at top of pipe

Element	FLOORS AND FINISHES Defect	Cause
Ground floors (a) Concrete	Damp penetration accompanied by detachment of finishings	Damp-proof membrane not provided, or damaged during construction, or not continuous with wall dpc
		Excessive ground water or springs exerting pressure on underside of slab and entering weak points in construction
		Temperature gradient – if the floor surface is colder than the ground underneath moisture will tend to rise and collect near the surface
	Arching and cracking of floor with possible displacement of adjoining external walls	Sulphate attack – soluble sulphates in unsuitable hardcore, e.g. colliery shale
	Cracking of slab	Excessive loading, or settlement on unconsolidated fill or made ground, or frost heave where slab exposed to the weather
(b) Timber	Decay evidenced by staining and splitting of floor boards with possibly a white, fluffy growth and characteristic musty smell	Dry rot caused by combination of following: (i) use of inadequately seasoned timber (ii) dampness – moisture content of timber should not exceed 20% (iii) lack of ventilation through omission or covering of air bricks

	FLOORS AND FINISHES	
Element	Defect	Cause
Upper floors (a) Concrete	Deflection	Use of shrinkable aggregates in concrete mix
		Drying shrinkage in an asymetrically reinforced slab
(b) Timber	Excessive springiness	Herringbone strutting omitted
		Joists weakened by notchings and holes for pipes and cables
		Joists affected by rot
		Shrinkage of secondary beams in framed floors resulting in tenons being pulled out of mortices in main beams
	Creaking	Boards not properly nailed down
	Decay in bathroom floors	Plumbing leaks and condensation, particularly where floors have an impervious covering and around base of WC
Floor finishes (a) Jointless		
Pitch mastic and asphalt	Indentations	Heavy furniture
	Surface deterioration and softening	Contact with grease, fats and oil
	Cracking and surface irregularities	Omission of separating membrane to isolate from movements of sub-floor
Epoxy resin	Surface deterioration	Excess of hardener in mix

Element	FLOORS AND FINISHES Defect	Cause
Floor finishes (a) Jointless (*contd.*)		
Magnesium oxychloride	Sweating of surface	Excess of magnesium chloride in mix
Granolithic	Cracking and lifting	Poor bond to base concrete – especially on old concrete which has not been thoroughly cleaned, hacked and wetted before laying finish
	Dusting of surface	Unsuitable materials in mix or poor workmanship
Terrazzo	Cracking and lifting	Poor key to base or omission of movement joints – floor should be divided into panels not exc. 1 m^2
(b) Sheet Linoleum	Irregular surface	Inadequate preparation of sub-floor
Flexible PVC	Lifting of edges and surface bubbles	Alkaline moisture from screed attacking adhesive
	Indelible brown stains on surface	Tar from adhesive diffusing through thickness of floor
	Scuff marks	Black rubber in footwear, caster tyres, etc.
Rubber	Surface deterioration	Contact with grease, fats and oil
(c) Tile and block Clay tiles		
	Arching or lifting of areas or rows of tiles	Drying shrinkage of screed, possibly accompanied by

	FLOORS AND FINISHES	
Element	Defect	Cause
Floor finishes (c) Tile and block Clay tiles (*contd.*)		moisture expansion of tiles Differential thermal movement of concrete base and tiles especially after a prolonged cold spell – concrete contracts two or three times as much as clay tiles
Thermo-plastic tiles	Surface deterioration	Softening of binders by solvent-based cleaners and polishes and by oil and grease
Vinyl tiles	Lifting of edges	Alkaline moisture from screed attacking adhesive
	Shrinkage of tiles causing wide gaps round edges	Movement of plasticiser from tile into adhesive causing adhesive to soften with result that tiles have tendency to move under traffic
	White strip around edges of tiles – 'window framing'	Soluble sulphates or sea salts in hardcore leaching through to surface
Cork tiles	Surface staining and general deterioration	Omission of sealing coat Contact with grease, fats and oil
Wood blocks	Staining and lifting of blocks	Rising damp causing moisture expansion of blocks Omission of expansion strip around perimeter of area to permit movement under varying conditions of temperature and humidity

Element	*WINDOWS, DOORS AND FITTINGS* Defect	Cause
External joinery	Decay of frames and sills evidenced by discoloration and lifting of paint film and 'cupping', cracking and softening of wood	Use of timber of low natural resistance to decay becoming sufficiently moist to allow wood-destroying fungi to grow (usually wet rot)
		Poor joint design and use of unsuitable glue allowing ingress of rainwater
		Inadequate weathering to horizontal surfaces not effectively shedding rainwater
		Dry bedding of glazing beads without spacers or back putty
		Lack of regular painting and reputtying allowing rainwater to enter open joints and cracks in putty
		Contact between unprimed frame and damp brickwork in absence of separating dpc or omission of cavity tray over head of opening
		Condensation running down inner face of glass and penetrating back putty, possibly aggravated by absence of condensation channel with adequate means of drainage therefrom
		Treatment prior to fixing, e.g. joinery exposed to weather in transit to site or stacked in open on site, coupled with breakdown of factory priming coat

WINDOWS, DOORS AND FITTINGS

Element	Defect	Cause
External joinery (*contd.*)	Water penetration and draughts under external doors	Worn threshold or absence of water bar and weather fillet
	Doors sticking	Warping of thin or insufficiently seasoned doors
		Settlement of foundations causing distortion of surrounding wall
Internal joinery	Pinholes in skirtings, frames, etc.	Beetle infestation
	Cracking and softening of skirtings, frames, etc.	Wet or dry rot
	Decayed window boards	Damp penetration under sill because of insufficient projection, lack of throating on underside of sill or omission of dpc under sill bridging the cavity
		Condensation – especially in bathrooms and kitchens
	Gaps at base of skirting	Shrinkage of floor joists, decay of wall plates or settlement of load-bearing walls
Stairs	Creaking	Shrinkage of wedges securing treads and risers to strings and breakdown of glued joint
		Omission of glued and screwed blockings between treads and risers or use of nails instead of screws

Element	WINDOWS, DOORS AND FITTINGS Defect	Cause
Stairs (*contd.*)	Gap between wall string and plaster	Settlement of flight or drying shrinkage of timber in new centrally heated property

Element	WALL AND CEILING FINISHES Defect	Cause
Plaster	Crazing of surface	Drying shrinkage
	Larger cracks	Differential movement of structural members
		Shrinkage of lightweight concrete partitions
		Settlement of foundations
		Vibration from external causes, e.g. heavy traffic
	Continuous crack at junction of ceiling and wall	Shrinkage
	Cracks in plaster at head of window frame	Differential shrinkage of wall and lintel when drying out
	Vertical crack in wall plaster	Differential shrinkage at point where application of plaster was stopped for a period or where a brick wall is finished flush against a concrete column
	Continuous straight cracks across ceiling	Deflection of weak, springy floor joists, or heavy loads in roof space on ceiling joists or joints between sheets of plasterboard which have not been scrimmed

	WALL AND CEILING FINISHES	
Element	Defect	Cause
Plaster (*contd.*)	Pitting or blowing in lime plaster	Imperfect slaking of lime
	Flaking of finishing coat	Inadequate key to undercoat either through unsuitable plaster for undercoat or undercoat not roughened
	Detachment of plaster from background	Background material not strong enough to resist movement of plaster
		Excessive absorption of water from plaster by porous background
		Inadequate key to background material such as smooth concrete with perhaps oil films from formwork
		Differential thermal movement of plaster and background
	Powdering	Improper setting and hardening of plaster attributable to unsound or stale materials or unsuitable mix
	Irregular and bulging ceiling in old properties	Plaster detached from laths either as a result of plaster keys broken by vibration or decayed laths
	Pattern staining of ceilings	Due to difference in thermal resistance of different parts of the ceiling. The cooler parts of the ceiling will tend to collect more dust than the warmer parts. Hence the ceiling immediately under

WALL AND CEILING FINISHES

Element	Defect	Cause
Plaster (*contd.*)		the joists being warmer than the surrounding surface will show as lighter strips
Wall tiles	Detachment and bulging of tiles on block partitions	Drying shrinkage of blocks, perhaps aggravated by moisture expansion of tiles in absence of suitable expansion joints to accommodate movement

DECORATIONS

Element	Defect	Cause
Paint films	Bleeding	Dyes or stains in certain paints or resin from knots bleeding through paint coats
	Blistering	Moisture entrapped under paint film vapourising under direct heat of sun
	Chalking	Powdery coating caused by disintegration of the binding material under the action of the weather or inferior quality paint
	Cracking (crazing)	Usually caused by ageing and ranges from fine hair cracks to crocodiling, a severe crazing resembling crocodile skin
	Efflorescence (patchy deposits on paint)	Soluble salts from the backing structure carried to the surface by water

	DECORATIONS	
Element	Defect	Cause
Paint films (*contd.*)		
	Fat edge	A ridge formed by applying fresh paint over a partly dried paint
	Flaking	Lifting of paint film caused by lack of adhesion to underlying surface, particularly smooth backgrounds such as glazed tiles
	Grinning through	Lack of opacity of paint film allowing underlying surface to show through
	Sagging	Downward movement of thick paint films prior to drying
	Sheeriness	Uneven gloss or glossy patches on flat or eggshell finishes due to uneven application giving different film thicknesses
	Splitting	Coloured spots or stains caused by mould growths
	Sticky film	Alkali attack by salts in backing
	Wrinkling (shrivelling)	Development of wrinkles during drying caused by applying paint too thickly
	Faint brown patches soon after first decoration	Salts from some types of bricks and clinker blocks leaching through gypsum plaster

Element	SERVICES Defect	Cause
Sanitary Plumbing (a) Pipework	Inadequate flow	Pipe sizes too small – possibly due to installation of additional fittings without increasing size of main supply pipe
		Insufficient head of water
		Obstruction in bore due to furring in hard water districts, or denting of pipe walls
	Bursts	Underground supply pipe too near surface or attack of unprotected steel pipes by corrosive soils
		Pipe in ventilated air space under floors, or close to eaves in roof space, or on external wall
		Joint failure due to poor workmanship, bimetallic action or, in the case of plastic, thermal movement of long lengths pulling pipe out of joint
	Signs of corrosion to external walls of pipe	Electrolytic action between copper and zinc in direct contact
		Dezincification – in acidic waters or alkaline waters of high chlorine content fittings of hot pressed (alpha-beta) brass lose zinc, leaving a soft porous copper which may extend through the wall of the fitting

Element	SERVICES Defect	Cause
Sanitary Plumbing (a) Pipework (*contd.*)	Noises	'Water hammer' in cold pipes caused by loose washer plates in valves
		'Knocking' in hot water pipes caused by obstructions in the flow and return pipes
		Other noises may be caused by loose or broken saddles or brackets or corroded screws and fixings
(b) Taps and stopcocks	Water leaks around top of spindle	Faulty gland packing
	Dripping	Faulty washer or worn valve seating
(c) Ball valves	Discharge from overflow	Valve fails to close properly due to faulty washer, or eroded seating or presence of grit or lime deposits
	Sticking in closed position	Usually follows a period of no use during which dirt and lime dry out in the working parts
	Inadequate flow	Use of high pressure instead of low pressure valve
	Hissing noise	High pressure feed
(d) Storage cistern	Rust pitting of galvanised steel cisterns	Electrolytic action due to iron filings from drilling holes left in tank or deposition of copper from water onto sides of tank

Element	SERVICES Defect	Cause
Sanitary Plumbing (d) Storage cistern (*contd.*)	Corrosion of fittings attached to cistern	Bimetallic action with fittings of brass or other copper bearing alloys
	Distortion of polythene cisterns	Inadequate base support
(e) Hot water cylinders	Corrosion of fittings attached to copper cylinders	Dezincification of brass fittings
Heating systems (a) Performance	Inadequate heat	Boiler too small or inadequate radiator surface
		Air lock or other obstruction to flow of hot water in system
		Thermostat faulty or incorrectly located
(b) Distribution pipes	Knocking and restricted flow of hot water	Air locking – entrainment of air in system through: (i) faulty design (ii) generation of gases by corrosion or bacteria (iii) air drawn into system through vent pipe which is not high enough to balance suction of pump (iv) use of pump at excessive pressure head
		Furring in hard water districts
(c) Radiators	Perforation of steel radiators	Entrained air causing corrosion
	Discoloured water from radiator vents	Mixing of primary and secondary circulation causing corrosion

Element	SERVICES Defect	Cause
Heating systems (c) Radiators (*contd.*)	Water leaking from glandless radiator valves	Faulty valve seating
(d) Boilers	Rust stains on face	Leak to joint of access plate
	Signs of leakage between sections	Faulty joints between sections or cracked section
	Water leaks from safety valve	Faulty valve seating, or weak main spring, or defective joint to boiler
(e) Circulating pump	Noise and vibration	Worn bearings
Ventilation (a) Ductwork	Noise	Transmission of noise from faulty fan
(b) Fans	Noise and vibration	Worn bearings
Electrical: Wiring	Failing to satisfy insulation resistance test	Deterioration and embrittlement of rubber sheathed cables perhaps accelerated by a humid environment
Drainage: Underground pipes	Blockage	Poor layout – pipe runs not in straight lengths between access points or laid to insufficient fall to clear soil
		Drain fractured by settlement, heavy traffic or tree roots
		Accumulation of builder's rubbish during construction
		Disposal of unsuitable materials via drains

Element	SERVICES Defect	Cause
Drainage: Underground pipes (*contd.*)		Accumulation of grease from kitchen waste
	Failing to satisfy tests (hydraulic or pneumatic)	Defective joints at: (i) base of soil pipe (ii) entry to manhole (iii) intermediate point
		Drain fractured as above
		Cracked flaunching in manholes

References

[1] BRE Digest 176 (1976) *Failure Patterns and Implications.* London: DOE, HMSO.

[2] BRE Digest (1974) *Building Defects and Maintenance.* Lancaster: The Construction Press.

[3] Eldridge, H.J. (1976) *Common Defects in Building.* London: DOE, HMSO.

Form of building inspection record

Reproduced from 'Practice in Property Management' (DOE, R & D Bulletin, 1970)

Building No...................

NOTE: 1. Insert 'S' in box if satisfactory or if there is a minor defect that can be left until next cyclic maintenance.
2. If a repair is required, describe briefly with rough estimate of cost.

BUILDING INSPECTION RECORD

Inspected by .... Date ....

ADDRESS   Next cyclic maintenance due
                                                 Year ....

1. *Roofs*   *EXTERNAL*

    1.1 Roof coverings

    1.2 Flashings and weatherings

    1.3 Verges

    1.4 Fascias

    1.5 Soffits

    1.6 Rafter feet (if exposed)

2. *Walls and Chimneys*

   2.1 Bricks

   2.2 Pointing

   2.3 Chimney pots

   2.4 Copings

   2.5 DPCs

   2.6 Rendering

   2.7 Decoration

3. *Doors and Windows*

   3.1 Wood

   3.2 Metal

   3.3 Glazing

   3.4 Putties or other seals

   3.5 Painting

4. *Plumbing*

   4.1 Soil and vent pipes

   4.2 RW gutters

   4.3 RW downpipes

   4.4 Painting

5. *Services and Ancillaries*

   5.1 Boundary walls

   5.2 Boundary fences

   5.3 Gates

5.4   Drives

5.5   Paths

5.6   Drainage

5.7   Cesspools

5.8   Septic tanks

5.9   Lighting fittings

5.10 Overhead wiring

5.11 Storage tanks

6. *Roof Space*   *INTERNAL*

6.1 Roof timbers

6.2 Chimney stacks and flues

6.3 Supply tanks

6.4 Plumbing

6.5 Insulation

6.6 Flooring

7. *1st Floor*

7.1   Ceilings

7.2   Walls

7.3   Fireplaces

7.4   Doors and windows

7.5   Floors

7.6   Decorations

7.7   Baths

7.8   L Basins

7.9   WCs

7.10 Glazed tiling

8. *Ground Floor*

8.1 Ceilings

8.2 Walls

8.3 Fireplaces

8.4 Doors and windows

8.5 Floors

8.6 Staircase

8.7 Decorations

8.8 Sinks

9. *General*

9.1 Central heating or DHW boiler

9.2 Central heating system

9.3 Electrical installation

9.4 Gas installation

9.5 Items not listed above

Form of classification of maintenance operations

Reproduced from 'Practice in Property Management' (DOE, R & D Bulletin, 1970)

PRIMARY CODE	SECONDARY CODE
1. EXTERNAL DECORATION	
2. INTERNAL DECORATION	
3. MAIN STRUCTURE	31. Foundations and Basements.
	32. Frame.
	33. External Walls, Chimneys and Flues (excluding Boiler Flues).
	34. External Windows and Doors including Glazing.
	35. Roof Structure.
	36. Roof Coverings.
	37. Roof Lights and Glazing.
	38. Gutters and Rainwater Pipes.
4. INTERNAL CONSTRUCTION	41. Ground Floors (where not part of Foundation Slab).
	42. Upper Floors.
	43. Staircases and Steps.
	44. Internal Walls and Partitions.
	45. Doors and Screens including Glazing.

PRIMARY CODE	SECONDARY CODE
5. FINISHES AND FITTINGS	51. Ceiling Finishes.
	52. Wall Finishes.
	53. Floor Finishes.
	54. Shelves, built-in Furniture and miscellaneous Joinery.
	55. Ironmongery.
	56. Cloakroom and similar miscellaneous Fittings.
6. PLUMBING AND SANITARY SERVICES	61. Cold Water Service Pipes, Storage Tanks, Cisterns and Valves.
	62. Hot Water Service Pipes, Storage Tanks and Cylinders, Domestic Boilers, Valves and Insulation.
	63. Sanitary Fittings including Taps and Traps.
	64. Waste, Soil and Vent Pipes.
7. MECHANICAL SERVICES INCLUDING HEATING AND VENTILATION AND GAS INSTALLATIONS	71. Boilers, Firing, Instrumentation and Automatic Controls, Flues.
	72. Steam and Hot Water Distribution including Heat Exchangers and Heating Appliances.
	73. Workshop Equipment, Lifting Appliances and Special Industrial Equipment.
	74. External Water Supply, Treatment and Storage Plant.
	75. Lifts.
	76. Air Conditioning, Ventilation and Refrigeration.
	77. General Utilities.
	78. Gas Installations and Equipment (except Kitchen Equipment).
8. ELECTRICAL SERVICES AND KITCHEN EQUIPMENT (ALL TYPES)	81. Electrical Generation and Prime Movers.
	82. Electrical Transmission and Distribution.
	83. Electrical Installations (Wiring Switch and Control Gear).
	84. Electrical Appliances and Fittings (except Kitchen Equipment).
	85. External Lighting and Airfield Lighting.

PRIMARY CODE	SECONDARY CODE
	86. Lightning Protection, ELV Systems and Equipment.
	87. Kitchen Equipment.
9. EXTERNAL AND CIVIL ENGINEERING WORKS	91. Roads, Car Parks, Parade Grounds, Hardstandings and Runways.
	92. Paths, Playgrounds and general Paved Areas.
	93. Fences, Gates and Boundary Walls.
	94. Drains and Ditches.
	95. Sewage Disposal.
	96. Water Storage Tanks and Reservoirs.
	97. Railway Tracks, Platform Docks, Wharves and Jetties, Sea Defence Walls.
	98. Horticultural and Arboricultural Works.
	99. Miscellaneous External Works.
0. MISCELLANEOUS AND ANCILLARY WORKS	01. Adaptations and Minor New Works.
	02. Replacement of Fire Damage.
	03. Routine Cleaning.

Agreement for minor building works

General

The current form was issued in 1980 by the Joint Contracts Tribunal which is constituted of representatives of the Royal Institute of British Architects, the Building Employers Confederation, the Royal Institution of Chartered Surveyors and various local authority and subcontractors' organisations. The form has been designed for use where the works are to be carried out for an agreed lump sum and where an architect/ supervising officer has been appointed on behalf of the employer.

The rubric to the form states that it is not suitable for use where bills of quantities have been prepared, or the employer wishes to nominate subcontractors, or the duration of the contract is such that full fluctuations provisions are needed or the works are of a complex nature or involve complex services. However, the form envisages the use of a priced specification or schedule for valuing variations and it would seem that there would be no bar to preparing a bill of quantities for use initially as an aid to tendering and then subsequently as a priced schedule. In this case the contractor should be advised to check the accuracy of the quantities since the bill will not be a contract document and any errors will not be subject to rectification. Also, although there are no provisions for nominating subcontractors it is anticipated that the contractor may wish to subcontract parts of the work. If the architect wishes a particular firm to undertake specialist parts of the work or to supply certain materials he may name the firm in the relevant contract document so that the contractor may obtain a quotation for inclusion in his tender. Such firms would then be on exactly the same footing as the other subcontractors and suppliers to the contractor. This should be done only where there are very good reasons for retaining control over who does the work or supplies the materials since, generally, it is more economic to leave the selection to the contractor. An alternative procedure that might be considered is that laid down in the JCT Standard Form of Building Contract, 1980 edition (the Standard Form) by which the contractor is required to select from three named firms.

The main factor determining the suitability of the form is undoubtedly the complexity of the work. If it is fairly straightforward traditional construction the form could be used for quite substantial jobs. Although the provisions do not spell out the procedures in such great detail as the Standard Form they provide an adequate framework for administering such contracts. The form has been extensively used since its introduction in 1968 and in the year 1978/79 sales of the form amounted to more than 50 per cent of all the other JCT forms put together, i.e. the Standard Forms for Private and Local Authority Use – With Quantities, Without Quantities and With Approximate Quantities. The Form starts with the basic Agreement followed by the Conditions and ends with a three-part Supplementary Memorandum.

Agreement

The agreement names the parties and identifies the work to be done by reference to the contract documents which may include one or more of the following – drawings, specification and schedules. The contract documents are to be signed by the parties and the contractor is required to price the specification or schedules or provide a schedule of rates for the valuation of variations. The articles state the contract sum, name the architect/supervising officer and provide for any disputes to be referred to arbitration at the request of either party. There is also provision for naming a quantity surveyor who, although he is not allocated any specific duties by the conditions, may be appointed by the employer to advise on the financial administration of the contract.

In the list of contents certain clauses are asterisked to indicate that they contain blanks which should be filled in by the parties or that the clause is optional and may be deleted as appropriate. The clauses are:

2.1   Commencement and completion.
       Dates to be inserted.
2.3   Damages for non-completion.
       Amount of liquidated damages to be stated.
2.5   Defects liability.
       Period stated if other than three months.
3.6   Variations.
       Indicate if priced specification, priced schedules or schedule of rates to be used for valuing variations.
4.1   Correction of inconsistencies.
       Strike out documents which are not contract documents.
4.2   Progress payments and retention.
       State retention percentage if other than 5 per cent.
4.3   Penultimate certificate.
       State percentage to be certified if other than 97½ per cent.
4.4   Final certificate.
       State period after practical completion within which contractor is to supply documentation for preparation of final account, if other than three months.
4.5   Contribution, levy and tax changes.
       Contractor to state percentage addition required on fluctuations.
5.4   Fair Wages Resolution (see page 395).
       Delete unless employer is a local authority.
5.5   Prevention of corruption.
       Delete unless employer is a local authority.
6.3A   Insurance of Works – Fire, etc. – New works.
       Delete if not new works. If retained fill in percentage for professional fees on amount insured.

6.3B Insurance of Works – Fire, etc. – Existing.
Delete if not an existing building.

Conditions

1.0 Intentions of the parties

Contractor's obligation

1.1 The Contractor shall with due diligence and in a good and workmanlike manner carry out and complete the Works in accordance with the Contract Documents using materials and workmanship of the quality and standards therein specified provided that where and to the extent that approval of the quality of materials or of the standards of workmanship is a matter for the opinion of the Architect/Supervising Officer such quality and standards shall be to the reasonable satisfaction of the Architect/Supervising Officer.

Architect's/Supervising Officer's duties

1.2 The Architect/Supervising Officer shall issue any further information necessary for the proper carrying out of the Works, issue all certificates and confirm all instructions in writing in accordance with these Conditions.

Commentary

1.1 '... with due diligence and in a good and workmanlike manner...' This is in accord with the obligations which would be implied by general law in the absence of express provision.

'... carry out and complete...' This imposes two separate obligations on the contractor: firstly, to carry out the works, and secondly, to complete. It should be noted that failure to complete would, in normal circumstances, constitute a breach of this fundamental obligation irrespective of the reason. In exceptional cases the contractor may obtain relief by the operation of the doctrine of frustration. This occurs where, without default of either party, some supervening event which could not have been foreseen at the time of entering into the contract renders further performance impossible. This is only available in very special circumstances and is rarely applicable to building contracts. It would not apply, for instance, to difficulties in carrying out the work in accordance with the drawings as these were capable of being foreseen at the time of entering into the contract.

'. . . in accordance with the Contract Documents . . .' These are the documents which have been so indicated in the Recitals and signed by the parties.

'. . . where . . . approval . . . is a matter for the opinion of the architect/supervising officer such quality and standards shall be to the reasonable satisfaction of the architect/supervising officer.' This applies only where the quality or standard has been specifically left to the architect's discretion, e.g. approval of facing bricks; in all other cases the contractor's obligation is to comply with the contract documents. Thus the architect cannot require a higher standard than specified nor can the contractor plead the architect's approval as an excuse for not complying with the specification.

1.2 '. . . shall issue . . .' The architect/supervising officer is under a duty to issue all information necessary for the execution of the works and failure to do so would be a breach of contract.

2.0 Commencement and completion

Commencement and completion
2.1 The Works may be commenced on

. . . .

and shall be completed by

. . . .

Extension of contract period
2.2 If it becomes apparent that the Works will not be completed by the date for completion inserted in clause 2.1 hereof (or any later date fixed in accordance with the provisions of this clause 2.2) for reasons beyond the control of the Contractor, then the Contractor shall so notify the Architect/Supervising

Officer who shall make, in writing, such extension of the time for completion as may be reasonable.

Damages for non-completion

2.3 If the Works are not completed by the completion date inserted in clause 2.1 hereof or by any later completion date fixed under clause 2.2. hereof then the Contractor shall pay to the Employer liquidated damages at the rate of £.......... per week for every week or part of a week during which the Works remain uncompleted.

Completion date

2.4 The Architect/Supervising Officer shall certify the date when in his opinion the Works have reached practical completion.

Defects liability

2.5 Any defects, excessive shrinkages or other faults which appear within three months [c] .... of the date of practical completion and are due to materials or workmanship not in accordance with the Contract or frost occurring before practical completion shall be made good by the Contractor entirely at his own cost unless the Architect/Supervising Officer shall otherwise instruct.

The Architect/Supervising Officer shall certify the date when in his opinion the Contractor's obligations under this clause 2.5 have been discharged.

[c] If a different period is required delete 'three months' and insert the appropriate period.

Commentary

2.1 '... may be commenced ...' The meaning is presumably that the works cannot be started before the stated date; a later start would be permitted provided the contractor meets the completion date.

2.2 '... for reasons beyond the control of the contractor ...' This is very broadly stated in that there are many matters outside the control of the contractor which may cause delay. It is advisable, therefore, to restrict the application of this provision by listing the events which, for this purpose, will be accepted as being outside the control of the contractor. This should be done by amending the printed conditions since the effect of clause 4.1 may be to nullify any provision in the specification which seeks to override, modify or affect in any way the conditions. Also, liquidated damages clauses which have not given an express power to grant an extension for delay caused by some act of the employer, e.g. ordering variations, have been held to be invalid. In Peak Construction (Liverpool) Ltd v. McKinney Foundations Ltd 1970 the court referred to the principle that the liquidated damages provisions of a contract cannot be enforced if there has been delay due to the employer's default and either the extension of time clause does not make provision for such delay, or there has been a failure to extend the time.

'. . . contractor shall so notify the architect/supervising officer . . .' This places a duty on the contractor to give notice if it becomes apparent that the works will not be completed on time for reasons beyond his control.

'. . . such extension . . . as may be reasonable.' Once notified of a valid cause of delay the architect/ supervising officer must grant a reasonable extension of time. This should be done as soon as practicable so that the contractor has a realistic completion date to which he can work.

2.3 '. . . liquidated damages . . .' The damages should be a genuine pre-estimate of the actual loss likely to be suffered as a result of the breach. If the amount is extravagant in relation to the greatest loss that could conceivably be suffered it may be construed as a penalty and be unenforceable.

2.4 '. . . practical completion.' This should not be construed as nearly complete, but as complete for all practical purposes, i.e. the value of any outstanding work, e.g. making good defects, is insignificant in relation to the total value of the contract.

2.5 '. . . defects . . . which appear within three months . . .' A different period may be substituted by the parties according to the nature and size of the contract. Also defects which appear after the end of the period are not covered by this provision but rectification costs could be claimed as damages for breach of contract.

'. . . made good by the contractor entirely at his own cost unless the architect/supervising officer shall otherwise instruct.' This could be interpreted either as an instruction forbidding the contractor to carry out the work or as an instruction relieving the contractor of his obligation to meet the costs of making good defects due to materials or workmanship not in accordance with the contract. The latter interpretation seems a little unlikely.

3.0 Control of the Works

Assignment
3.1 Neither the Employer nor the Contractor shall, without the written consent of the other, assign this Contract.

Subcontracting
3.2 The Contractor shall not subcontract the Works or any part thereof without the written consent of the Architect/ Supervising Officer whose consent shall not unreasonably be withheld.

Contractor's representative
3.3 The Contractor shall at all reasonable times keep upon the Works a competent person in charge and any instructions given to him by the Architect/Supervising Officer shall be deemed to have been issued to the Contractor.

Exclusion from the Works
3.4 The Architect/Supervising Officer may (but not unreasonably or vexatiously) issue instructions requiring the exclusion from the Works of any person employed thereon.

Architect's/Supervising Officer's instructions
3.5 The Architect/Supervising Officer may issue written instructions which the Contractor shall forthwith carry out. If instructions are given orally they shall, in two days, be confirmed in writing by the Architect/Supervising Officer.

If within seven days after receipt of a written notice from the

Architect/Supervising Officer requiring compliance with an instruction the Contractor does not comply therewith then the Employer may employ and pay other persons to carry out the work and all costs incurred thereby may be deducted by him from any monies due or to become due to the Contractor under this Contract or shall be recoverable from the Contractor by the Employer as a debt.

Variations
3.6 The Architect/Supervising Officer may, without invalidating the contract, order an addition to or omission from or other change in the Works or the order or period in which they are to be carried out and any such instruction shall be valued by the Architect/Supervising Officer on a fair and reasonable basis, using where relevant prices in the priced specification/schedules/schedule of rates [d].

Instead of the valuation referred to above, the price may be agreed between the Architect/Supervising Officer and the Contractor prior to the Contractor carrying out any such instruction.

Provisional sums
3.7 The Architect/Supervising Officer shall issue instructions as to the expenditure of any provisional sums and such instructions shall be valued in accordance with clause 3.6 hereof.

Commentary

3.1 The burden of a contract cannot in any case be assigned without the consent of the other party although it is usually permissible to assign the benefits, e.g. it is not uncommon for a contractor to assign money due on a contract in order to secure an advance from a bank.

3.2 If the contractor wishes to subcontract part of the work the consent of the architect/supervising officer should not be unreasonably withheld and may be challenged by arbitration.

3.3 '. . . competent person in charge . . .' The qualifications of the person in charge should be consistent with the nature and scope of the work.

3.5 The architect/supervising officer is given an apparently unrestricted power to issue instructions with which the contractor must forthwith comply. Oral instructions are to be confirmed 'in two days' (presumably this means within two days of giving the instruction) and, therefore, it could be argued that the contractor would be safe in acting on oral instructions in that failure to confirm would be a breach on the part of the architect/supervising officer. Compliance with instructions may be enforced by giving written notice to the contractor and if he fails to comply within seven days of receipt of the notice the employer may engage other persons to do the work and recover the cost from the contractor.

3.6   It is interesting to note that instructions may be issued not only in respect of additions and omissions but also in respect of the order or period in which the Works are to be carried out. It is doubtful whether this would extend to directing the contractor to do the work in a particular way. The general position is that it is the function and right of the contractor to carry out the building operations as he thinks fit subject to any constraints which are laid down in the contract documents and for which he would have allowed in his tender. Also, the power to alter the period of the work, which must mean either lengthening or shortening the period, is additional to the provision for extending the time in the event of delay. However, such changes are to be valued on a fair and reasonable basis using where relevant the prices contained in the contract documents and could include an allowance for any loss or expense suffered by the contractor as a result of having to do the work in a different order. For cost control purposes it would be better to adopt the alternative of agreeing inclusive prices with the contractor before the work is carried out.

3.7   Provisional sums for work which cannot be entirely foreseen at the time of tendering are to be expended as instructed by the architect/supervising officer. It should be noted that there is now no provision for P.c sums in respect of nominated subcontractors or nominated suppliers and where it is desired that a particular firm should do part of the work or supply certain materials the firm should be named in the contract documents.

4.0 Payment

Correction of inconsistencies

4.1 Any inconsistency in or between the Contract Drawings [d] and the Contract Specification [d] and the schedules [d] shall be corrected and any such correction which results in an addition, omission or other change shall be treated as a variation under clause 3.6 hereof. Nothing contained in the Contract Drawings [d] or the Contract Specification [d] or the schedules [d] shall override, modify or affect in any way whatsoever the application or interpretation of that which is contained in these Conditions.

Progress payments and retention

4.2 The Architect/Supervising Officer shall if requested by the Contractor, at intervals of not less than four weeks calculated from the date for commencement, certify progress payments to the Contractor in respect of the value of the Works properly executed, including any amounts either ascertained or agreed under clauses 3.6 and 3.7 hereof, and the value of any materials and goods which have been reasonably and properly brought upon the site for the purpose of the Works and which are adequately stored and protected against the weather and other casualties less a retention of 5 per cent/......per cent [e] and less any previous payments made by the Employer, and the Employer shall pay to the Contractor the amount so certified within 14 days of the date of the certificate.

Penultimate certificate

4.3 The Architect/Supervising Officer shall within 14 days after the date of practical completion certified under clause 2.4 hereof certify payment to the Contractor of $97\frac{1}{2}$ per cent/..... per cent [f] of the total amount to be paid to the Contractor under this contract so far as that amount is ascertainable at the date of practical completion, including any amounts either ascertained or agreed under clauses 3.6 and 3.7 hereof less the amount of any progress payments previously made by the Employer, and the Employer shall pay to the Contractor the amount so certified within 14 days of that certificate.

Building maintenance management

Final certificate

4.4 The Contractor shall supply within three months/
.... [g] from the date of practical completion all documentation reasonably required for the computation of the amount to be finally certified by the
Architect/Supervising Officer and the Architect/Supervising
Officer shall within 28 days of receipt of such documentation,
provided that the Architect/Supervising Officer has issued the
certificate under clause 2.5 hereof, issue a final certificate
certifying the amount remaining due to the Contractor or due
to the Employer as the case may be and such sum shall as
from the fourteenth day after the date of the final certificate
be a debt payable as the case may be by the Employer to the
Contractor or by the Contractor to the Employer.

Contribution, levy and tax changes [h]

4.5 Contribution, levy and tax changes shall be dealt with by the
application of Part A of the Supplementary Memorandum to
the Agreement for Minor Building Works. The percentage
addition under Part A, clause A5 is ...... per cent.

Fixed price

4.6 No account shall be taken in any payment to the Contractor
under this Contract of any change in the cost to the Contractor of the labour, materials, plant and other resources employed in carrying out the Works except as provided in clause
4.5 hereof, if applicable.

[d] Delete as appropriate to follow any deletions in the recitals on page 1.
[e] If a different retention is required delete '5 per cent' and insert the appropriate
percentage.
[f] The alternative should be completed where a percentage other than 5 per cent
has been inserted in clause 4.2.
[g] If a different period is required delete 'three months' and insert the appropriate
period.
[h] Delete clause 4.5 if the contract period is of such limited duration as to make the
provisions of part A of the Supplementary Memorandum to this agreement
inapplicable.

Commentary

4.1 It should be noted that only those corrections of
inconsistencies which involve a change are to be treated
as variations. Also that the provision that nothing in
the contract documents 'shall override, modify or affect

in any way whatsoever . . . these Conditions' would invalidate any statements in the specification or other documents which are not in accord with the printed conditions. This is contrary to the normal legal presumption that words written specifically for a particular contract take precedence over standard conditions.

4.2 Progress payments to be made monthly 'if requested by the contractor'. Note that materials are only included if they have been reasonably and properly brought upon the site, i.e. not prematurely, and if they are adequately stored and protected. There is no provision for including materials stored in the contractor's yard even though they may have been prepared specifically for the works.

4.3 It is recognised that full documentation from the contractor may not be available at the time of preparing the penultimate certificate and therefore the amount is described as being so far as is ascertainable at the date of completion. However, the $2\frac{1}{2}$ per cent retained should cover any small inaccuracies caused by lack of information.

4.4 This places an obligation on the contractor to produce all documentation reasonably necessary to determine the final amount. The final certificate is to be issued within 28 days of receipt of the documentation or after the issue of the certificate of making good defects, whichever is the later. Thus, the final certificate could not be issued earlier than the end of the defects liability period even where no documentation is required. It also provides for any overpayments on the penultimate certificate to be recovered as a debt payable by the contractor to the employer.

4.5 Price fluctuations are limited to changes in the types and rates of contribution, levy and tax payable by the contractor in respect of his employees and on the types and rates of duty and tax payable on materials, electricity and, where specifically stated in the contract documents, fuels. If there is a change in the type or rate of tax the net amount of the increase or decrease is added to or deducted from the contract sum. The detailed provisions are given in Part A of the Supplementary

Memorandum and are similar to those contained in
Clause 38 of the Standard Form. If full fluctuations are
required, i.e. increases or decreases in wage rates and
materials prices as well as statutory contributions and
taxes, Clause 39 of the Standard Form (which is published separately) could be used. This would, however,
require the deletion of 4.6.

5.0 Statutory obligations

Statutory obligations, notices, fees and charges

5.1 The Contractor shall comply with, and give all notices required by, any statute, any statutory instrument, rule or order
or any regulation or byelaw applicable to the Works
(hereinafter called 'the statutory requirements') and shall
pay all fees and charges in respect of the Works legally recoverable from him. If the Contractor finds any divergence
between the statutory requirements and the contract documents or between the statutory requirements and any instruction of the Architect/Supervising Officer he shall immediately
give to the Architect/Supervising Officer a written notice
specifying the divergence. Subject to this latter obligation, the
Contractor shall not be liable to the Employer under this
Contract if the Works do not comply with the statutory requirements where and to the extent that such non-compliance
of the Works results from the Contractor having carried out
work in accordance with the Contract Documents or any instruction of the Architect/Supervising Officer.

Value Added Tax

5.2 The sum or sums due to the Contractor under clause 2 hereof
of this Agreement shall be exclusive of any value added tax
and the Employer shall pay to the Contractor any value
added tax properly chargeable by the Commissioners of
Customs and Excise on the supply to the Employer of any
goods and services by the Contractor under this Contract
in the manner set out in Part B of the Supplementary
Memorandum to the Agreement for Minor Building Works.

Statutory tax deduction scheme

5.3 Where at the date of tender the Employer was a 'contractor', or where at any time up to the issue and payment of the final certificate the Employer becomes a 'contractor', for the purposes of the statutory tax deduction scheme referred to in Part C of the Supplementary Memorandum to the Agreement for Minor Building Works, Part C of that Memorandum shall be operated.

Fair Wages Resolution

5.4 [i] This relates to a Resolution passed by the House of Commons in 1946. The Fair Wages Resolution was rescinded by the government in 1982 but has been left in the contract because a requirement to comply may still remain in the Standing Orders of some local authorities.

Prevention of corruption

5.5 [i] The Employer shall be entitled to cancel this contract and to recover from the Contractor the amount of any loss resulting from such cancellation, if the Contractor shall have offered or given or agreed to give to any person any gift or consideration of any kind or if the Contractor shall have committed any offence under the Prevention of Corruption Acts, 1889 to 1916, or shall have given any fee or reward the receipt of which is an offence under sub-section (2) of section 117 of the Local Government Act, 1972, or any re-enactment thereof.

[i] To be deleted unless the Employer is a local authority.

Commentary

5.1 This provision is somewhat ambiguous in that it starts by imposing a liability on the contractor to comply with all statutory requirements and regulations and later, subject to a proviso that 'if' he finds any divergence between the statutory requirements and the contract documents he shall immediately give written notice to the architect/supervising officer, exempts him from liability to the employer to the extent that non-compliance results from following the contract documents.

5.2 The detailed provisions relating to Value Added Tax are set out in Part B of the Supplementary Memorandum.

Briefly they provide that the architect/supervising officer is to advise the contractor of amounts certified for interim payment and that the contractor is to give an immediate written assessment of the values of materials and services which will be chargeable and the respective rates of VAT. Upon receipt of the assessment the employer shall calculate the total VAT chargeable and add it to the amount of the interim payment.

On the issue of the certificate of making good defects the contractor is to prepare a written final statement of the values of goods and services chargeable, the rate(s) of VAT applicable and the total amount of VAT already received by him. The employer is then required to pay any balance due to the contractor within 28 days of receipt of the statement or, in the event of overpayment, the contractor is to reimburse the employer within a similar period. If the employer disagrees with the final statement issued by the contractor he may require the contractor to obtain the decision of the Commissioners. If the employer wishes the contractor to appeal against the decision of the Commissioners he must secure him against all costs and other expenses involved. Upon receipt of any VAT properly paid the contractor is to issue to the employer an authenticated receipt in the form of those issued by the Joint Contracts Tribunal.

It should be noted that services supplied to the final client for new construction are currently zero rated. The Finance Act 1984 abolished the tax distinction between repairs and alterations, and both are now subject to VAT. Thus all work to an existing building is standard rated whether it is described as alteration, improvement, reconstruction or repair. Generally to qualify as a new building the work should be physically separate from any existing building.

5.3 The detailed provisions are given in Part C of the Supplementary Memorandum. The object of the Finance (No. 2) Act, 1975, was to exercise control over the so called 'lump' workers by requiring payments made by 'contractors' to 'subcontractors' to be subject to deduction for income tax unless the payee has a valid current tax certificate issued by the Inland Revenue. For this purpose the employer is a contractor if:

(*a*) He carries on a business which includes construction operations. Note, that this includes businesses which maintain their own permanent building department, irrespective of size, whose activities amount in whole or in part to 'construction operations'.

(*b*) He is a local authority, development corporation or new town commission or a housing corporation or association.

The tax certificates 714 I, P or C for individuals, partnerships and companies respectively are to be submitted to the employer (the 'contractor') by the 'subcontractor' not later than 21 days before the first payment is due, or alternatively, the 'subcontractor' is to inform the employer in writing that he is not entitled to be paid without the statutory deduction. In the latter case the subcontractor is to state not later than seven days before each future payment the amount which represents the cost of materials so that the employer can assess the tax liability on the balance of labour costs.

5.4 } These clauses are only required where the employer is a
5.5 } local authority.

6.0 Injury, damage and insurance

Injury to or death of persons

6.1 The Contractor shall be liable for and shall indemnify the Employer against any expense, liability, loss, claim or proceedings whatsoever arising under any statute or at common law in respect of personal injury to or death of any person whomsoever arising out of or in the course of or caused by the carrying out of the Works, unless due to any act or neglect of the Employer or of any person for whom the Employer is responsible. Without prejudice to his liability to indemnify the Employer the Contractor shall maintain and shall cause any subcontractor to maintain such insurances as are necessary to cover the liability of the Contractor or, as the case may be, of such subcontractor, in respect of personal injury or death arising out of or in the course of or caused by the carrying out of the Works. Provided that nothing in this clause contained shall impose any liability on the subcontractor in respect of negligence or breach of duty on the part of the Employer, the Contractor, his other subcontractors or their respective servants or agents.

Damage to property

6.2 The Contractor shall, subject to clause 6.3A/B, be liable for and indemnify the Employer against and insure and cause any subcontractor to insure against any expense, liability, loss, claim of proceedings in respect of any damage whatsoever to any property real or personal insofar as such damage arises out of or in the course of or by reason of the carrying out of the Works and is due to any negligence, omission or default of the Contractor or any person for whom the Contractor is responsible or of any subcontractor or person for whom the subcontractor is responsible.

Insurance of the Works – Fire, etc. – New Works [j]

6.3A The Contractor shall in the joint names of Employer and Contractor insure against loss and damage by fire, lightning, explosion, storm, tempest, flood, bursting or overflowing of water tanks, apparatus or pipes, earthquake, aircraft and other aerial devices or articles dropped therefrom, riot and civil commotion, for the full value thereof plus ...... per cent [k] to cover professional fees all work executed and all unfixed materials and goods intended for, delivered to, placed on or adjacent to the Works and intended therefor (except temporary buildings, plant, tools and equipment owned or hired by the Contractor and his subcontractors).

Upon acceptance of any claim under the insurance mentioned in this clause 6.3A the Contractor shall with due diligence restore or replace work or materials damaged and dispose of any debris and proceed with and complete the Works. The Contractor shall not be entitled to any payment in respect of work other than the monies received under the said insurance (less the percentage to cover professional fees) and such monies shall be paid to the Contractor under certificates of the Architect/Supervising Officer at the periods stated in clause 4.0 hereof.

Insurance of the Works – Fire, etc. – Existing structures [j]

6.3B The Works (and the existing structures together with the contents thereof owned by him and for which he is responsible) and all unfixed materials and goods intended for, delivered to, placed on or adjacent to the Works and intended therefor (except temporary buildings, plant, tools and equipment owned or hired by the Contractor or any subcontractor) shall be at the sole risk of the Employer as regards loss or damage by fire, lightning, explosion, storm, tempest, flood, bursting or overflowing of water tanks, apparatus or pipes, earthquake, aircraft and other aerial devices or articles dropped therefrom, riot and civil commotion, and the Employer shall maintain adequate insurance against that risk.

If any loss or damage as referred to in this clause occurs then the Architect/Supervising Officer shall issue instructions for the reinstatement and making good of such loss or damage in accordance with clause 3.5 hereof.

Evidence of insurance

6.4 The Contractor shall produce, and shall cause any subcontractor to produce, such evidence as the Employer may reasonably require that the insurances referred to in clauses 6.1 and 6.2 and, where applicable 6.3A, hereof have been taken out and are in force at all material times. Where clause 6.3B hereof is applicable the Employer shall produce such evidence as the Contractor may reasonably require that the insurance referred to therein has been taken out and is in force at all material times.

[j] Delete 6.3A or 6.3B if not applicable.
[k] Percentage to be inserted.

Commentary

6.1 '. . . personal injury to or the death of any person . . .'
The contractor accepts full liability and indemnifies the
employer against losses and claims in respect of personal injury and death *unless due to any act or neglect of
the employer.* The contractor is required to maintain
such insurances as are necessary to cover his own liability in this respect and ensure that subcontractors
do likewise. No distinction is made between injury to
the contractor's employees and third parties although
separate policies would apply. Under the Employer's
Liability (Compulsory Insurance) Act, 1969, it is compulsory for the employer, i.e. the contractor in this
context, to take out insurance against liability for bodily
injury or disease sustained by his employees in the
course of employment of at least £2 million in respect
of claims from any one instance. The third party or
public liability insurance is not compulsory by law
and is usually for a lesser sum.

6.2 '. . . any damage . . . to any property real or personal.'
In this case the contractor accepts liability *only in so far
as the damage is caused by any negligence, omission or
default on his part.* It should be noted that the contractor is only required to insure against his own liability and that it would not cover the employer if there
had been no negligence on the contractor's part. Thus
in the case of Gold *v.* Patman and Fotheringham Ltd
1958 damage was caused to adjoining property through
no negligence of the contractor. In accordance with the
terms of the contract the contractor had merely insured
his own liability and the employer was unable to claim
on the insurance or seek a remedy from the contractor
on the indemnity since there had been no negligence on
the part of the contractor. The simplest solution would
be to have a joint names policy to cover all third-party
liabilities irrespective of who is negligent.

The amount of insurance cover which the contractor is
required to take out should be related to the proximity,
nature and use of adjoining premises and the extent of
possible consequential losses. It is important not to

confuse the size of the job with the potential size of a
loss which could give rise to a claim.

6.3A This applies to new works where the contractor would
be in control of the site. The list of insured risks
although fairly extensive does not mention such
common risks as theft and vandalism and an 'all risks'
policy would be better. Also payments for restoring
damaged work are to be made on certificates of the
architect/supervising officer although it is the contractor who actually holds the policy.

6.3B '. . . and the existing structures together with the contents thereof owned by him and for which he is responsible.' This applies where work is being carried to
an existing building which would normally already be
covered by insurance. In such cases it would be necessary to check the existing policy to see if there is a
'workmen allowed' clause permitting the carrying out
of repairs, alterations, etc. without prejudicing the
policy or a 'non-invalidation' clause confirming that the
policy will not be invalidated by any act or omission
which increases the risk for a short time and which is
beyond the control or knowledge of the insured. If in
any doubt it is prudent to inform the insurance company before the work is started.

'. . . shall issue instructions for the reinstatement . . .'
Whether or not reinstatement is desirable will depend
upon the extent of the damage. Clearly, if the damage is
very extensive the restoration work may be of a scale
and character which is quite beyond the resources of
the contractor. Oddly there is no provision for determination in this situation but as the contract would
become incapable of execution it would be frustrated.
Where the damage is less extensive the instruction
would relate to redoing the original work, possibly including some element of improvement to be valued on a
fair and reasonable basis.

6.4 The actual policies should be checked and in particular
any exclusion clauses. Also 'Act only' policies should be
viewed with suspicion. One of the provisions of the
Finance Act No. 2 (1975) requires subcontractors who
apply for a 714 certificate to show evidence of having

third party insurance covering bodily injury with a limit of indemnity of £250 000. In order to reduce the premium some insurers issue 'Act only' policies but these cover only bodily injuries and do not cover damage to property. Also the limit is an aggregate and not in respect of each and every loss.

7.0 Determination

Determination by Employer

7.1 The Employer may but not unreasonably or vexatiously by notice by registered post or recorded delivery to the Contractor forthwith determine the employment of the Contractor under this Contract if the Contractor shall make default in any one or more of the following respects:

1. if the Contractor without reasonable cause fails to proceed diligently with the Works or wholly suspends the carrying out of the Works before completion;
2. if the Contractor becomes bankrupt or makes any composition or arrangement with his creditors or has a winding up order made or (except for the purposes of reconstruction) a resolution for voluntary winding up passed or a receiver or manager of his business or undertaking is duly appointed or possession is taken by or on behalf of any creditor of any property the subject of a charge.

In the event of the Employer determining the employment of the Contractor as aforesaid the Contractor shall immediately give up possession of the site of the Works and the Employer shall not be bound to make any further payment to the Contractor until after completion of the Works. Provided always that the right of determination shall be without prejudice to any other rights or remedies which the Employer may possess.

Determination by Contractor

7.2 The Contractor may but not unreasonably or vexatiously by notice by registered post or recorded delivery to the Employer forthwith determine the employment of the Contractor under this Contract if the Employer shall make default in any one or more of the following respects:

1. if the Employer fails to make any progress payment due

under the provisions of Clause 4.3 hereof within 14 days of such payment being due;

2. if the Employer or any person for whom he is responsible interferes with or obstructs the carrying out of the Works or fails to make the premises available for the Contractor in accordance with clause 2.1 hereof;

3. if the Employer suspends the carrying out of the Works for a continuous period of at least one month;

4. if the Employer becomes bankrupt or makes a composition or arrangement with his creditors, or has a winding up order made or a resolution for voluntary winding up passed or a receiver or manager of his business is appointed or possession is taken by or on behalf of any creditor of any property the subject of a charge.

Provided that the employment of the Contractor shall not be determined under clauses 7.2.1, 7.2.2 or 7.2.3 hereof unless the Employer has continued the default for seven days after receipt by registered post or recorded delivery of a notice from the Contractor specifying such default.

In the event of the Contractor determining the Employment of the Contractor as aforesaid the Employer shall pay to the Contractor, after taking into account amounts previously paid, such sum as shall be fair and reasonable for the value of work begun and executed, materials on site and the removal of all temporary buildings, plant, tools and equipment. Provided always that the right of determination shall be without prejudice to any other rights or remedies which the Contractor may possess.

Commentary

7.1 This clause gives the employer the right to determine immediately the employment of the contractor if he defaults in one of the two ways stated. Unlike the Standard Form it does not provide for a preliminary warning notice before the notice of determination.

'. . . not unreasonably or vexatiously . . .' This imposes a limitation on the employer's right to determine the employment of the contractor. What is reasonable will depend upon the surrounding circumstances and if unreasonable may constitute wrongful dismissal and entitle the contractor to damages.

'... determine the employment of the contractor ...'
It should be noted that it is the contractor's employment that is determined and not the contract. Certain
clauses will subsist after the determination of the
contractor's employment. Thus any dispute over the
notice of determination may be referred to arbitration.
'... without reasonable cause fails to proceed
diligently ...' Any dispute concerning the reasonableness of the cause or the diligence with which the work is
being carried out may be referred to arbitration.
'... wholly suspends the carrying out of the
Works ...' The complete abandonment of the work
before substantial completion without lawful cause
would amount to repudiation which, quite apart from
the contract conditions, would entitle the employer to
rescind the contract and sue for damages.
'... if the contractor becomes bankrupt ...' In the
absence of express provisions in the contract a trustee
in bankruptcy or a liquidator has the right to disclaim
onerous contracts or to continue with the work if this is
in the best interests of the creditors. This clause gives
the employer the initiative to decide whether or not to
determine the contractor's employment. If the employer wishes to take advantage of the clause he should
not delay in giving the requisite notice, otherwise it may
be said that he has waived his rights under the contract.
'... without prejudice to any other rights or
remedies ...' The employer is not precluded from treat-
ing the breach as repudiation and suing for damages.
The contract does not provide for the assessment and
recovery of losses suffered by the employer as a result
of the breach.

7.2   This clause gives the contractor similar rights to determine his own employment if the employer defaults in
any of the specified ways.
'... fails to make any progress payment ...' Payment
is due within 14 days of the date of the certificate and if
not made within a further 14 days the contractor is
entitled to exercise his right under this clause.
'... interferes with, or obstructs the carrying out of
the Works or fails to make the premises available ...'

Failure to make the premises available would amount to a fundamental breach which would entitle the contractor to treat the contract as at an end and claim damages. This may be preferable to the contractual remedy which does not go beyond payment for work begun and executed, materials and removal of plant, etc.; a remedy which would give scant compensation if the contractor has not been given access to the site.

'. . . if the Employer becomes bankrupt . . .' In general a contractor is not bound to work on credit for an employer if the latter becomes bankrupt.

Unlike the employer, the contractor is required to give notice specifying the default and wait seven days before determining his employment.

8.0 Supplementary Memorandum

Meaning of references in 4.5, 5.2 and 5.3
8.1 The references in clauses 4.5, 5.2 and 5.3 to the Supplementary Memorandum to the Agreement for Minor Building Works are to that issued for use with this Form by the Joint Contracts Tribunal as endorsed hereon.

Commentary

The Supplementary Memorandum is divided into three parts:

Part A – Contribution, Levy and Tax Changes (clause 4.5)
Part B – Value Added Tax (clause 5.2)
Part C – Statutory Tax Deduction Scheme (clause 5.3)

The provisons are outlined under the relevant clauses.

Index